3rd International Conference on Applied Geophysics (AG 2017)

E3S Web of Conferences Volume 24 (2017)

Gniew, Poland
21 - 23 June 2017

Editors:

Pilecki Zenon
Marcak Henryk

ISBN: 978-1-5108-5228-0

Printed from e-media with permission by:

Curran Associates, Inc.
57 Morehouse Lane
Red Hook, NY 12571

Some format issues inherent in the e-media version may also appear in this print version.

Printed by Curran Associates, Inc. (2017)

For additional information, please contact EDP Sciences – Web of Conferences at the address below.

EDP Sciences – Web of Conferences
17, Avenue du Hoggar
Parc d'Activité de Courtabœuf
BP 112
F-91944 Les Ulis Cedex A
France

Phone: +33 (0) 1 69 18 75 75
Fax: +33 (0) 1 69 28 84 91

contact-edps@webofconferences.org

Additional copies of this publication are available from:

Curran Associates, Inc.
57 Morehouse Lane
Red Hook, NY 12571 USA
Phone: 845-758-0400
Fax: 845-758-2633
Email: curran@proceedings.com
Web: www.proceedings.com

TABLE OF CONTENTS

Statement of Peer review

In submitting conference proceedings to E3S Web of Conferences, I certify to the Publisher that I adhere to the **Policy on Publishing Integrity** of the journal in order to safeguard good scientific practice in publishing.

1. All articles have been subjected to peer review administered by the proceedings editors.
2. Reviews have been conducted by expert referees, who have been requested to provide unbiased and constructive comments aimed, whenever possible, at improving the work.
3. Proceedings editors have taken all reasonable steps to ensure the quality of the materials they publish and their decision to accept or reject a paper for publication has been based only on the merits of the work and the relevance to the journal.

Title, date and place of the conference

AG 2017 - 3rd International Conference on Applied Geophysics (Gniew, Poland, June 21-23, 2017)

Proceedings editor(s):

Pilecki Zenon, Ass. Prof.
Marcak Henryk, Prof.

Date and editor's signature

24/10/2017

17, avenue du Hoggar - PA de Courtabœuf – BP 112 - 91944 Les Ulis Cedex A (France)
Tél. : 33 (0)1 69 18 75 75 - Fax : 33(0)1 69 07 45 17 – www.edpsciences.org

DOI: 10.1051/e3sconf/20172401001

Basic principles for the identification of landslides using geophysical methods

Zenon Pilecki[1*]

[1]Mineral and Energy Economy Research Institute, Polish Academy of Sciences, 31-261 Krakow, Poland

Abstract. Cases of damage to structures in landslide areas in Poland, and sometimes construction catastrophes, occur most frequently in periods of intense precipitation (e.g. in 1997, 2000, 2002 and 2010). The greatest landslide risk occurs in mountainous areas in the Flysch Carpathians, and the number of landslides posing major problems exceeds 20,000. Terrain surface relief and geological structure including the tectonics, properties of the geological medium, presence of excessive water mostly due to intense precipitation, and human activity have an impact on landslide risk. This paper is aimed at presenting basic principles for the geophysical identification of landslides gained from experience in various regions of Poland, particularly in the flysch formations in the Carpathian Mountains. The geophysical methods which are of the greatest importance in the study of landslides are specified. Their advantages and limitations are presented. The principles of geophysical methods applied at the stage of designing and implementing research, processing and interpreting data, analysis of results, and their documentation are discussed.

1 Introduction

Recognition of landslide risk has been the subject of much research work. Cruden and Varnes [1], Dikau et al. [2], and Roy [3] described interesting presentations of landslide processes and landslide classifications. Similar issues concerning the conditions of landslides in Poland were presented by Kleczkowski [4], Bober [5], Thiel [6], Zabuski et al. [7], Bober and Thiel [8], Rybicki et al. [9], Wiłun [10], Borecka et al. [11].

Landslide areas are difficult to use for construction purposes [12]. In Poland, cases of damage to buildings occur, sometimes involving catastrophic impacts on structures, most frequently in periods of intense precipitation (e.g. in 1997, 2000, 2002 and 2010). The greatest risk of landslides occurs in mountainous areas in the Flysch Carpathians, and the number of landslides posing major problems exceeds 20,000 [13, 14] (Fig. 1). The existence of such a large number of recorded landslides in this area is influenced by the terrain surface relief and the geological structure including tectonics, the properties of the geological medium, the presence of excessive water mostly due to intense precipitation, and human activities. There are occasional earthquakes in the Beskid Żywiecki and in the Pieniny Mountains which also influence the activation of landslides [13].

[*] Corresponding author: pilecki@meeri.pl

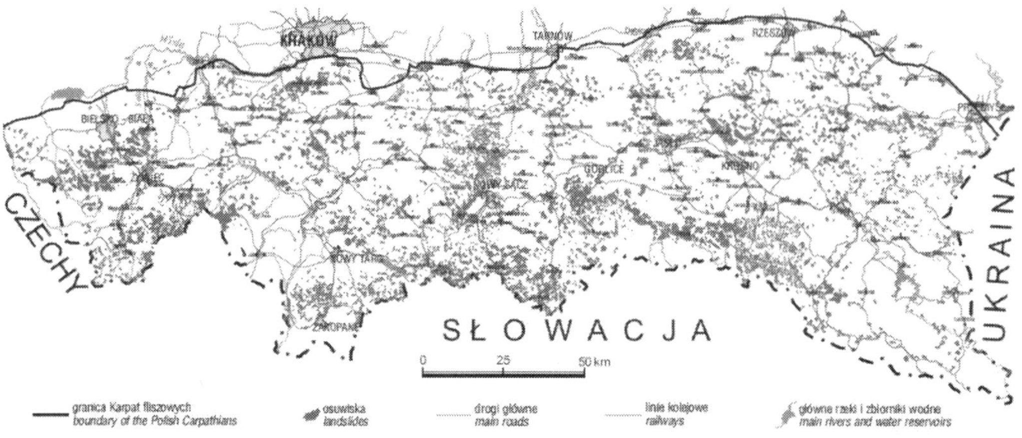

Fig. 1. Map of landslides in the Polish Flysch Carpathians compiled from data from the Polish Geological Institute updated to the end of 2000 [14].

Under the conditions found in the Flysch Carpathians, it should be emphasised that there is a scientific and practical need to develop a geophysical method for identifying the risk of landslides because:

1. The method of recognising landslide processes under the conditions in the Flysch Carpathians is primarily by geodetic, inclinometric and piezometric measurements. In the last two decades, geophysical methods have become more and more widespread, but their use has mainly been experimental.
2. The fast development of geophysical equipment and specialised software enables 3D/4D geophysical models, more attractive for engineering purposes, to be developed. These results, however, require proper presentation and analysis in a manner that can be understood by the construction engineer.
3. Geophysical information can be helpful at every stage of making construction decisions.

Jongmans and Garambois [15] presented an interesting diagnosis of the relatively limited use of geophysical methods in identifying landslide hazards. They found that *geophysical methods provide images in terms of physical parameters, which are not directly linked to the geological and mechanical properties required by geologists and engineers. Another reason comes from a tendency among some geophysicists to overestimate the quality and reliability of the results.* Undoubtedly, these statements are relevant to much geophysical work, so it is important to continually emphasise the principles behind proper identification of hazards using geophysical means.

This paper is aimed at presenting the basic principles of geophysical identification of landslides. It specifies the geophysical methods which are of greatest importance in the study of landslides. Their advantages and limitations are presented. The principles of the geophysical methods applied at the stage of designing and carrying out measurements, processing and interpreting data, analysis of results and their documentation are described.

2 Geophysical methods used for landslide identification

Usually the identification of landslides by geophysical methods is presented worldwide in terms of the results of measurements on various types of landslides. Jongmans and Garambois [15] classified these applications in terms of two general objectives:

1. Locating the boundaries of geophysical contrast within the geological medium affected by slide processes including the failure surfaces. Four main different possibilities are considered :

- The geophysical contrasts are of lithological origin (layering or pre-slide weathering), and the failure surface mainly coincides with a geological interface or layer. These features range from local fracture zones to larger regional structural features. Joints, fractures and fault zones may be dry, fluid-filled or filled with clays or weathered rock. The detectability of these features increases with the size of the feature and with the presence of distinctive pore fluids or conductive fill material.
- Geophysical contrasts are also controlled by lithological variations, but the failure surface cuts the structure in a more complex way and may or may not be deduced from the geophysical image depending on the landslide activity, the heterogeneity of the geological medium and the resolution of the technique.
- The failure surface or presumed failure surface is mainly detected by geophysical propagation methods.
- Geophysical contrast arises between the slide and the unaffected mass arising from the cumulative or separate actions of displacement, weathering and an increase in water content.

2. Recognition of water content within the geological medium affected by slide processes, for which electrical and electromagnetic methods were usually applied.

The identification of landslides using geophysical methods has been the subject of many publications that deal with this issue in a more general way, e.g. Bogoslovsky and Ogilvy [16], Mc Cann and Foster [17], Caris et al. [18], Hack [19], and Jongmans and Garambois [15]. It should be emphasised that the approach by which this issue is presented in these publications is different due to the intensive development of equipment and the processing and interpretation of geophysical data over the last few decades.

In the specific conditions of the flysch formations in the Polish section of the Carpathians, geophysical methods are most frequently used in the identification of such geological and hydrogeological elements as:

- Courses of probable slip surfaces, which can also be seen as the border between contrasting physical properties in the geological medium. These boundaries usually occur at the contact of a less weathered rock basement with a more weathered upper part, most often affected by landslide processes.
- Geophysical boundaries, including lithological or larger fractures meeting the criterion of contrasting properties in a geological medium.
- The location of the ground water table and the impact of waterlogging on the formation of the slip surface.
- Weak zones characterised by unfavourable spatial variations of physical properties in the geological medium, such as waterlogged zones, intense fracturing, slide mass affected by landslide processes, etc.

Landslides are one of the most difficult geophysical issues due to their many causes and the complex nature of the landslide process. Under geological-engineering conditions in relation to typical landslides in Poland, the most important geophysical methods used are:

- Ground Penetrating Radar (GPR) in the 2D/3D/4D profiling scheme,
- Seismic survey via 2D/3D/4D refraction profiling, seismic tomography, multichannel surface analysis (MASW),
- Electrical Resistivity Profiling (ERP) in the 2D/3D/4D profiling scheme, Vertical Electrical Sounding (VES), and Electrical Resistivity Tomography (ERT).

Examples of such research are: Bestyński and Trojan [20], Ślusarczyk [21], Dziewański and Pilecki [22], Karczewski and Ziętek [23], Pilecki et al. [24, 25], Bednarczyk [26], Ostrowski et al. [27], Kamiński et al. [28], Harba and Pilecki [29].

These methods were also highlighted as the most important ones used in recognising the structure and properties of soils and rocks by the American Association for Testing and

Materials [30], reapproved in 2011 [31]. Other examples are the use of geophysical methods such as: reflection seismics, seismic interferometry, borehole tomography methods, and magnetic or self-potential methods. There have also been great developments in remote sensing techniques over the last dozen years or so.

3 Basic advantages and disadvantages of geophysical methods

Geophysical methods are used for non-invasive investigation of the properties and structure of a geological medium and the physical processes taking place within it. The non-invasive nature of this investigation indicates that there is no need for direct access to the geological medium, e.g. by borehole or other excavation. Depending on the kind of geophysical method, different properties of the geological medium are investigated: electromagnetic, elastic, electric, magnetic, physical-mechanical and other. In order to have the possibility of recognising borders with distinct properties, contrast is an important element in geophysical investigations.

Geophysical methods have their advantages and limitations [32]. The main advantages of these methods are as follows:

- They are **efficient**, as they allow for relatively large areas to be surveyed relatively quickly and continuously in comparison with, for example, local (one point) boreholes. Geophysical studies should be financially more attractive than a set of boreholes or other excavations.
- They are **non-invasive**, because they do not require invasive access to the geological medium. They are carried out on the terrain surface or via excavation, if possible.
- They are **multi-informative**, as they use different physical fields. Therefore it is possible to describe the geological medium in a variety of complementary types of information. If one performs a series of measurements on the same profile at different intervals of time, it is possible to investigate the development of the landslide process.
- The **penetration depth** varies depending on the geophysical method used and its methodology. Geophysical surveys carried out for landslide recognition are directed towards identifying the geological medium to a depth of ten metres or so and, if necessary, to a few tens of metres, which is generally sufficient to design a protective structure for the landslide.
- **Resolution**, as in the case of penetration depth, this varies depending on the geophysical method used and its methodology. For the basic geophysical methods: GPR, seismic, and electrical methods, greater resolution results in less depth of penetration.

The main disadvantages of geophysical methods are as follows:

- The **ambiguity of the results of geophysical interpretations** requires them to be correlated with geological-engineering information, or with the results from other geophysical measurements. At least two complementary geophysical methods are suggested.
- **Geophysical recordings are sensitive** to various types of disturbance during measurement.
- **Geophysical images are of limited resolution**, particularly with increased distance from the source (transmitting antenna, vibration source or current electrode). In the case of active geophysical methods there is a compromise between resolution and depth to which identification is attempted. In the case of passive methods there is ambiguity in the identification of features because an

object at a particular depth can generate a similar anomaly to a larger object at a greater depth.

- The **time required for processing and interpretation of the geophysical data** varies and is often difficult to determine. It depends on the type of geophysical method used and the measurement conditions. For more complicated tasks, this time can be extended, resulting in higher costs.

It is worth underlining that the theoretical principles behind geophysical methods are quantitative, however when actually applied, they often yield interpretations that are qualitative [31]. Some geophysical methods provide data from which a preliminary interpretation can be made in the field, for example GPR. Some methods require that the data should be processed before any quantitative interpretation can be made, for example, seismic refraction.

Geophysical methods can be used in two situations in relation to landslide identification: (1) in an initial investigation of the landslide hazard (2) to confirm the anomalies provided by other measurements or to provide additional information.

Geophysical analysis must be conducted by a competent professional to ensure that it is consistent with geological-engineering conditions.

In summary, it should be emphasised that geophysical methods provide special information that cannot be obtained by other methods. Jongmans and Garambois [15] stated that *almost all the advantages of geophysical methods correspond to the disadvantages of geotechnical techniques and vice versa.*

4 Basic principles of geophysical identification of landslides (on the basis of [32])

Appropriate methodologies for geophysical measurement and for processing and interpretation of geophysical data should be properly designed to achieve the most informative results.

At the stage of design and realisation of geophysical measurements, the following principles are suggested:

a) Initially identify the research tasks involved, including the scope of the research and collect available geological-engineering and geophysical information about the study area. Site inspection is also required to assess the measurement conditions.
b) Specify the type of geophysical methods which are most effective in solving the research task involved when considered the conditions under which measurements will take place. One should select geophysical methods in which the cost of acquiring geophysical information is cost-effective compared to other methods. In designing the measurement methodology, it is important to note that:
   - ✓ The choice of geophysical method depends on the ability to distinguish different types of borders and weak zones in the geological medium.
   - ✓ The results of the measurement should be validated with geological-engineering information coming from, for example, an archival borehole. If necessary, a borehole should be made for data comparison.
   - ✓ A method should be selected that is not susceptible to measurement disturbances and which cannot be reduced satisfactorily at the processing stage. In difficult conditions of strong measurement disturbances, it is advisable to perform test measurements together with test processing and interpretation of the data recorded.
   - ✓ To solve the research task using geophysical methods one should try to apply at least two methods based on different properties of the medium being examined. The application of one geophysical method is associated with a high risk of

ambiguity in analysis. This ambiguity can be reduced by taking into account the results available from archival investigations, including data from different excavations and the results from using other methods, during the processing and interpretation process.

c) Design an interpretation methodology in accordance with the measurement conditions and determine how to document the results of the investigation.
d) Carry out the geophysical investigation project including the methodology of measurement, processing and interpretation together with the presentation of final results. The project must be submitted to the contractor.
e) When carrying out geophysical surveys, measurement profiles or measuring points should be determined by the geodetic method. Sufficient accuracy in identifying the location of the measuring point is achieved by means of GPS and measuring tape. General information about the measurements (date, profile number, location, etc.), regardless of their recording in the memory of the measuring equipment, should be written in the operator's notebook.

The following principles are suggested at the geophysical data processing and interpretation stage, during analysis of the final results, as well in their documentation:

a) Field registrations should be organised and archived in the computerised database in clearly-described directories. These data can be used, for example, for reinterpretation.
b) Data processing should primarily aim at improving the signal-to-noise ratio.
c) Data interpretation, usually associated with inverse procedures used for the description of the geological medium with the help of mathematical models, should match (satisfy) the field data. The measure of the quality of the calculations should be the fit of the theoretical model to the field data.
d) Visualisation of the results of the calculations should be legible and understandable for other interested professionals.
e) Comprehensive analysis of the results of the investigation should be made taking into account all the available information, using their own and archival data. On the basis of the analysis, conclusions are drawn from studies that are relevant to further decision making, including engineering for the design of structures to provide protection against landslides. An important part of the analysis is information about the errors involved in and limitations of the geophysical method used.
f) The documentation should be understandable to the construction engineer or other specialist. The description of the measurement methodology and interpretation methodology should be presented in such a way that the investigation can be repeated on the basis of this description.

5 Summary

This paper presents the basic principles for the identification of landslides in Poland by geophysical methods, particularly in the flysch formations of the Polish Carpathian Mountains.

In the specific conditions of the Flysch Carpathians, geophysical methods are most frequently used for the identification of the presumed slip surface, lithological boundaries or larger fractures fulfilling the criterion of contrast in physical properties of the geological medium, the location of the groundwater table and the impact of waterlogging on the formation of slip surfaces, and the location of weak zones characterised by unfavourable spatial variations in the properties of the medium, e.g. zones of waterlogging, intensive fracturing, zones affected by slide processes, etc. The recognition and analysis of all these problems by traditional geotechnical methods is often impossible.

It is stressed in the paper, that appropriate methodologies of geophysical measurement and for the processing and interpretation of the geophysical data should be properly designed to achieve the most informative results.

In the last two decades, the rapid development of geophysical equipment and specialised software has enabled the presentation of ever more attractive 2D geophysical models, and increasingly 3D and 4D models. These models are increasingly attractive for engineering purposes, particularly in time-lapse mode.

This paper also emphasises that each geophysical method also has its advantages and limitations, which have to be provided for in assessing their reliability in site identification.

References

1. D.M. Cruden, D.J. Varnes, *Landslide Types and Processes*, Special Report, Transportation Research Board, National Academy of Sciences **247**, 36-75 (1996)
2. R. Dikau, D. Brundsen, L. Schrott, M.-L. Ibsen, *Landslide recognition: identification, movement and causes* (Wiley, Chichester, UK, 1996)
3. R. C. Sidle, H. Ochiai, *Landslides: processes, prediction, and land use*, Water Resources Monograph **18** (American Geophysical Union, Washington, 2006)
4. A. Kleczkowski, *Osuwiska i zjawiska pokrewne* (Wyd. Geologiczne, Warsaw, 1995)
5. L. Bober, *Rejony osuwiskowe w polskich Karpatach fliszowych i ich związek z budowa geologiczną regionu*, Biul. Inst. Geologicznego **340**, XXIII, 115-162 (1984)
6. K. Thiel (editor), *Właściwości fizyko-mechaniczne i modele masywów skalnych polskich Karpat fliszowych*, Hydrotechnika **19** (Publ. Institute of Hydroengineering PAS, Gdańsk, 1995)
7. L. Zabuski, L. Bober, K. Thiel, *Osuwiska we fliszu Karpat polskich* (Publ. Institute of Hydroengineering PAS, Gdańsk, 1999)
8. L. Bober, K. Thiel, L. Zabuski, *Zjawiska osuwiskowe w polskich Karpatach fliszowych. Geologiczno-inżynierskie właściwości wybranych osuwisk* (Publ. Institute of Hydroengineering PAS, Gdańsk, 1997)
9. S. Rybicki, W. Margielewski, A. Domagała, *Landslide on Palenica Mt near Szczawnica (Pieniny Klippen Belt) and their connection with extrema rains in July 1997*, Przegląd Geologiczny **46**, 11, 1162-1170 (1999)
10. Z. Wiłun, *Zarys geotechniki* (Wyd. Komunikacji i Łączności, Warsaw, 2000)
11. A. Borecka, M. Cała, J. Flisiak, M. Kowalski, P. Krokoszyński, J. Maciaszek, Z. Pilecki, S. Rybicki, H. Woźniak, *Metody badania, monitorowania i stabilizacji osuwisk* (AGH University of Science and Technology Press, to be published)
12. Z. Pilecki, J. Kłosiński, E. Pilecka, J. Karczewski, J. Ziętek, *Wpływ zawodnienia osuwiska na rozpoznanie granic jego nieciągłości metodami falowymi*, Gospod. Surowcami Min. **24**, 2/3, 427-444 (2008)
13. W. Rączkowski, *Osuwiska polskich Karpat fliszowych*, 259-262 (Przew. 72 Zjazdu Pol. Tow. Geol. MAEGS-12, Kraków, 2001)
14. D. Poprawa, W. Rączkowski, *Osuwiska Karpat*, Przegląd Geologiczny **51**, 8, 685-692 (2003)
15. D. Jongmans, S. Garambois, B. Soc. Geol. Fr. 178, 2, 101-112, (2007) doi: 10.2113/gssgfbull.178.2.101
16. V.A. Bogoslovsky, A.A. Ogilvy, Geophysics, 42, 562–571 (1997) doi: 10.1190/1.1440727

17. D.M. Mc Cann, A. Forster, Eng. Geol., 29, 59–78 (1990) doi: 10.1016/0013-7952(90)90082-C
18. J.P.T. Caris, Th.W.J. Van Asch, Eng. Geol., 31, 249-276 (1991) doi: 10.1016/0013-7952(1)90011-9
19. R. Hack, Surv. Geophys. 21, 423–448 (2000) doi: 10.1023/A:1006797126800
20. Z. Bestyński, J. Trojan, *Metody geofizyczne w badaniach stateczności zboczy skalnych*, Mat. Badawcze seria specjalna nr 4 (Institute of Meteorology and Water Management – National Research Institute, Warsaw, 1975)
21. R. Ślusarczyk, *Możliwości zastosowania geofizyki inżynierskiej w problematyce budownictwa lądowego i wodnego*, 109-124 (Geofizyka w inżynierii i ochronie środowiska, Dębe near Warsaw, 2001)
22. J. Dziewański, Z. Pilecki, *Ocena warunków geologiczno-inżynierskich na terenie powierzchniowych ruchów masowych na przykładzie osuwiska w Zgłobicach*, Studia, Rozprawy, Monografie **109**, (Publ. Mineral and Energy Economy Research Institute PAS, Krakow, 2002)
23. J. Karczewski, J. Ziętek, *Wykorzystanie metody georadarowej w geofizyce środowiska*, Publ. of the Inst. of Geoph. Polish Academy of Science, M-27, 352, 223-232 (2002)
24. Z. Pilecki, J. Ziętek, J. Karczewski, E. Pilecka, J. Kłosiński, *The effectiveness of recognizing of failure surface of the Carpathian flysch landslide using wave methods*, Proc. 13th European Meeting of Environmental and Engineering Geophysics **P48**, 1-5 (Istanbul, 03 September 2007)
25. Z. Pilecki (editor), E. Popiołek, J. Karczewski, J. Ziętek, J. Kłosiński, A. Baranowski, E. Pilecka, Ł. Ortyl, J. Pszonka, K. Krawiec, *Wpływ rozdzielczości metod falowych na efektywność rozpoznania granic nieciągłości osuwiska* (Publ. Agencja Wydawniczo-Poligraficzna „ART-TEKST", Krakow, 2008)
26. Z. Bednarczyk, *Ground Penetrating Radar (GPR) scanning in geological and geotechnical recognition of mountain site*, 731-738 (22 SAGEEP Symposium, Fort Worth, Texas USA, EEGS, US Army Research Office, 2009)
27. S. Ostrowski, B. Rybak-Ostrowska, M. Lasocki, *Wykorzystanie przypowierzchniowych badań geofizycznych w rozpoznaniu budowy geologicznej na przykładzie stref osuwiskowych w Karpatach*, Przegląd Geologiczny **61**, 1, 67-73 (2013)
28. M. Kamiński, P. Zientara, M. Krawczyk, *Wykorzystanie tomografii elektrooporowej do badania osuwisk – osuwisko Dzianisz (południowo-zachodnie Podhale)*. Przegląd Geologiczny **62**, 4, 198-2003 (2014)
29. P. Harba, Z. Pilecki, Landslides **14**, 3, 1225-1233, doi:10.1007/s10346-016-0779-2 (2016)
30. ASTM, *Standard Guide for Selecting Surface Geophysical Methods, Designation D-6429* (Philadelphia, 1999)
31. ASTM, *Standard guide for selecting surface geophysical methods, Designation: D6429-99*, American Society for Testing Materials (Philadelphia, 2011)
32. Z. Pilecki, E. Pilecka, *Podstawowe zasady stosowania metod geofizycznych w badaniu osuwisk i terenów zagrożonych osuwiskami*, Zeszyty Naukowe – Bulletin of the Mineral Energy Economy Research Institute of the Polish Academy of Sciences **92**, 131–140 (2016)

Numerical modelling of GPR electromagnetic fields for locating burial sites

José M. Carcione[1], *Jerzy* Karczewski[2], *Ewelina* Mazurkiewicz[2], *Ryszard* Tadeusiewicz[2], and *Sylwia* Tomecka-Suchoń[2*]

[1]Istituto Nazionale di Oceanografia e di Geofisica Sperimentale OGS, 34010 Trieste, Italy
[2]AGH University of Science and Technology, 30-059 Krakow, Poland

Abstract. Ground-penetrating radar (GPR) is commonly used for locating burial sites. In this article, we acquired radargrams at a site where a domestic pig cadaver was buried. The measurements were conducted with the ProEx System GPR manufactured by the Swedish company Mala Geoscience with an antenna of 500MHz. The event corresponding to the pig can be clearly seen in the measurements. In order to improve the interpretation, the electromagnetic field is compared to numerical simulations computed with the pseudo-spectral Fourier method. A geological model has been defined on the basis of assumed electromagnetic properties (permittivity, conductivity and magnetic permeability). The results, when compared with the GPR measurements, show a dissimilar amplitude behaviour, with a stronger reflection event from the bottom of the pit. We have therefore performed another simulation by decreasing the electrical conductivity of the body very close to that of air. The comparison improved, showing more reflections, which could be an indication that the body contains air or has been degraded to a certain extent that the electrical resistivity has greatly increased.

1 Introduction

In this work, the possibility of locating burial sites on the basis of GPR radargrams is analysed. An extensive introduction to the method can be found, among others, in the following works: [1 - 5]. Carcione [6,7], illustrate the method and the related theory for detecting contaminated sites and coffins, respectively. The search of burial sites with the GPR was described by Doolittle and Bellantoni [8] and Schultz [9], among others. Apart from the GPR technique, other geophysical methods can be used to locate graves, for instance, electrical methods [10, 11] and magnetometry [12].

The target here is a domestic pig cadaver (Sus Scofa Domestica) buried in south-eastern Poland. The pit bottom is located at approximately 1m depth, and for the purposes of modelling, the cadaver cross-section dimensions are assumed to be 1m x 0.5m. Clay and loamy sand are predominant in the place of the measurement. A set of GPR profiles were

*Corresponding author: tomecka@agh.edu.pl

acquired, the aim of which is to assess the suitability of the technique at burial sites with difficult geological conditions (clay and loamy sand with high attenuation of electromagnetic waves).

In specific cases, in order to solve ambiguity during GPR interpretation, numerical modelling of the electromagnetic wave field was used. In the paper by Golebiowski [13] numerical modelling was applied for proper interpretation of radargrams recorded in river dike; in the paper by Golebiowski [14] the synthetic radargrams were used to improve the interpretation of GPR data recorded in archaeological site; in paper by Marcak and Golebiowski [15] numerical modelling of two-phase flow and electromagnetic wave field were applied for proper interpretation of radargrams recorded in hydrocarbon contaminated sites. In the paper a modelling algorithm based on the pseudo-spectral Fourier method has been employed and the results compared with the radargram recorded by a ProEx System GPR with a 500MHz shielded antenna.

Electromagnetic numerical simulation in heterogeneous media is essential to validate the geological interpretations. The GPR wave modelling is, in general, one-dimensional in the literature. Goodman and Nishimura [16] proposed a two-dimensional wave simulation method based on ray-tracing techniques. However, this approach suffers the disadvantages of ray methods, i.e., the impossibility of modelling the full wave field at all frequency ranges (e.g., the complete set of multiples and diffractions), and the generation of non-uniform dissipative waves at material interfaces (e.g., [17]). It is important to model the correct frequency dependence of the permittivity. At radar frequencies (≈50 MHz - 1 GHz), various dielectric dispersion processes occur. In moist soils, the most important are ionic conductivity and relaxation effects. The relaxation produces an increase in attenuation with frequency, since the molecules begin to lag the applied field and increase the real effective conductivity. This phenomenon is well described by a Debye relaxation peak, having its analogy in the Zener rheological model (also called standard linear solid) used in visco elasticity (e.g., [17]). Moreover, at high frequencies, the response of free charges may lag the electric field and produce an out-of-phase component, contributing to the real effective permittivity. These phenomena, i.e., dielectric relaxation and out-of phase electric currents, are introduced into the Maxwell equations by means of time-domain permittivity and conductivity functions [6, 7].

2 Field study

As mentioned above, the target is a domestic pig cadaver buried at a depth of 1m. The pit was made in a ground with predominance of clay and loamy sands. Figure 1 shows a simplified scheme. Figure 2 shows pig carcass prepared for burial.

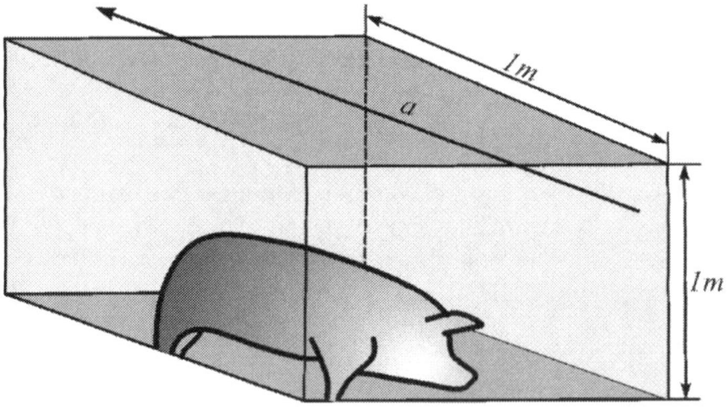

Fig. 1. Sketch of buried pig with an indicated direction of profiling.

Fig. 2. The pig cadaver prepared for burial.

The measurements have been performed with a ProEx System GPR. Antennas of various designs and frequency ranges were used, from 100 to 800MHz. In this work, only the results for the 500MHz antenna are presented. On the basis of resolution arguments, it can be shown that an antenna of such frequency is the optimal one for searching burial sites located at the depth of 2m or less. A set of radar profiles of a length of approximately 10m have been acquired. The data has been processed with the ReflexW software manufactured by the Sandmeier company.

3 Modelling the electromagnetic field

Figure 3 shows the model with a dimension of 2 x 1.5 m. The ellipse, with semi-axis of 0.5m and 0.25m, represents the pig cadaver.

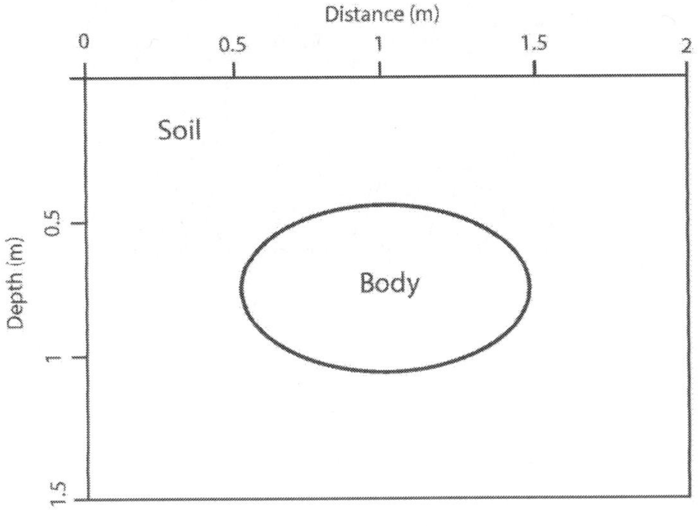

Fig. 3. Model representing a cross-section of a pig buried in soil.

The electromagnetic properties are given in Table 1, where ε_r and μ_r are the relative permittivity and magnetic permeability, respectively, and σ is the electrical conductivity [18].

Table 1. Model electromagnetic properties.

	ε_r	σ(S/m)	μ_r
soil	16	0.04	1
body	50	0.8	1

The pseudo-spectral Fourier method is used to obtain the synthetic electromagnetic field (radargrams). A detailed introduction to this method can be found in Carcione [6, 7, 19] and Carcione et al. [20]. The model is discretized into 0.02m cells with 135 x 135 grid points. The model is extended 0.35m on the sides, top and bottom in order to include absorbing-boundary conditions (suppression of the electromagnetic wave at the model boundaries to avoid wraparound). The simulations have a source time history applied to the horizontal electric current (a Ricker wavelet) with a central frequency of 500MHz.

4 Results and analysis

The field measurements were conducted with the ProEx georadar of the Swedish company MALA GeoScience. A set of profiles were acquired in the testing area, with the use of various antennas. In Figure 4, a radargram corresponding to the 500 MHz antenna is displayed. During the field measurements the following parameters were assumed: number of samples: 730; number of stacks: 8; distance intervals Δx=0.02m.

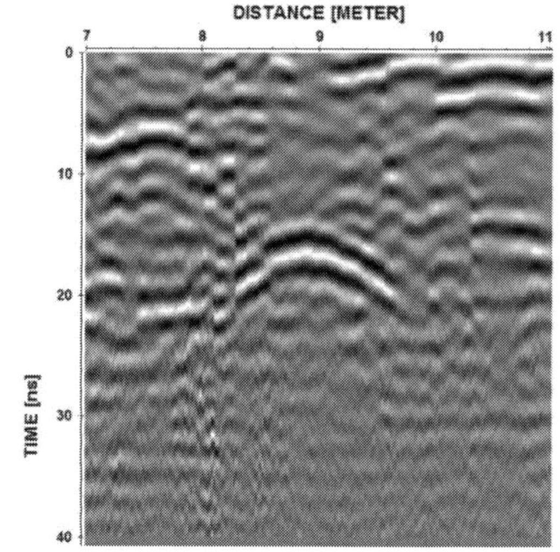

Fig. 4. Radargram recorded with the 500 MHz antenna along the profile a (Figure 1).

The radargram presented in the further part of the article was subjected to standard digital signal processing in the ReflexW software of the German company SandmeierGeo (among the other procedures were applied: t_0 correction, background removal, subtracking average). As can be appreciated, there are events between 10 and 15 ns (the top of the body presumably) and a characteristic hyperbola at approximately 22 ns between 8.5 and 9.5 m horizontal distance, where the bottom of the pit is located. Somehow, these events correspond to the pig cadaver.

We have modelled the response with the above mentioned numerical algorithm. The radargrams obtained with the pseudo-spectral method are shown in Figure 5. Snapshots (not shown) indicate that a complicated interference pattern is produced by reverberations inside the body. The high dissimilar permittivity constants of the soil and the body (at 500MHz) generates the strong reflection event, which originates at the top of body. However, the dissipation due to the high conductivity inside the body has damped the reflection coming from the bottom. As can be seen in Figure 5, there is practically no energy transmission below the body, contrary to the to the real case, where there is a strong event related to the bottom of the pit. This discrepancy may indicate that the wave loss inside the body is not so strong, meaning that the conductivity can be much lower, due to the presence of air or degradation of the medium. Then, we have performed another simulation with air replacing the pig. As is well known, air has a negligible conductivity, in practice a very high resistivity. Figure 6 shows the corresponding radargrams, where it can be seen that more events are present below the main hyperbola with apex at 12 ns.

Further work may consider other interpretation tools. For example, Mazurkiewicz et al. [21] use artificial neuron networks to process data. The Multilayer Perceptron network is used, which is based on an empirically sized hidden layer, trained by wave propagation. The fundamental difficulty associated with the method of introducing information on the analysed GPR image is overcome by substituting the signal samples values (that can be many) with a parametric description obtained with a polynomial approximation [22]. These methods will be applied in a future work dealing with the present problem.

5 Conclusions

We have acquired GPR profiles above a pit containing a pig cadaver. The method shows that the surface radar can help detect the body, especially in simple geological condition (similar to described in article). To support the interpretation, we have performed numerical simulations with a full-wave direct modelling technique, based on the Fourier pseudo-spectral method. In order to obtain a good qualitative comparison order to enhance the reflections from the bottom of the pit. A too high conductivity implies an energy loss due to wave dissipation (due to ionic currents) and those events can hardly be seen. Low conductivity, on the contrary, implies a series of events from top to bottom of the pit, which show a better comparison with the real data.

This work was financed within the agreement numbers 11.11.140.645 and 15.11.140.474 (AGH University of Science and Technology, Faculty of Geology, Geophysics and Environmental Protection).

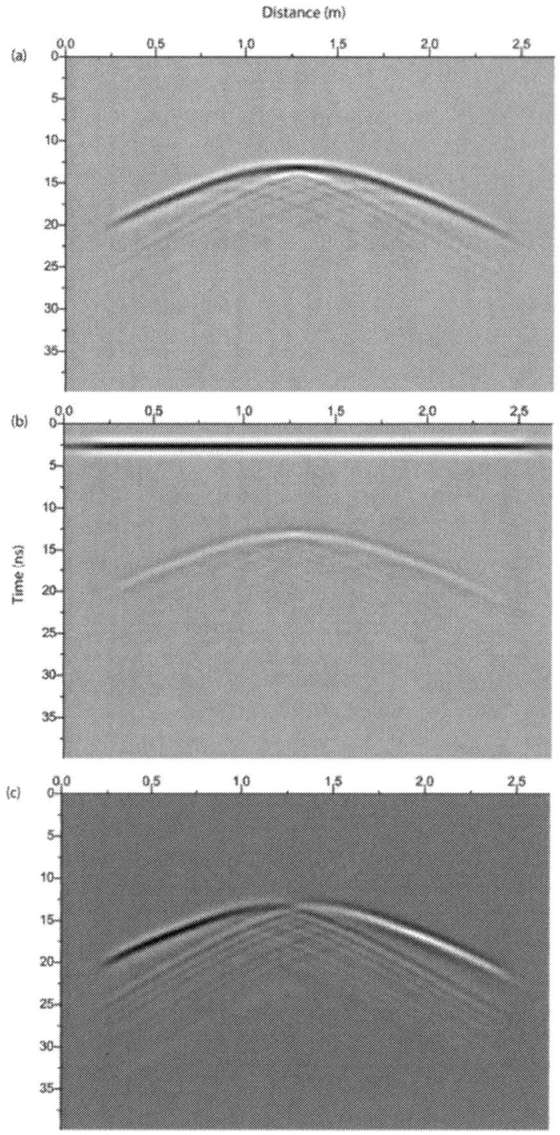

Fig. 5. Synthetic radargrams representing the magnetic field H_y (a) and the horizontal and vertical electric fields E_x (b) and E_z (c). The frequency of the radar is 500MHz (pseudo-spectral Fourier method, see Carcione (1996 a, b).

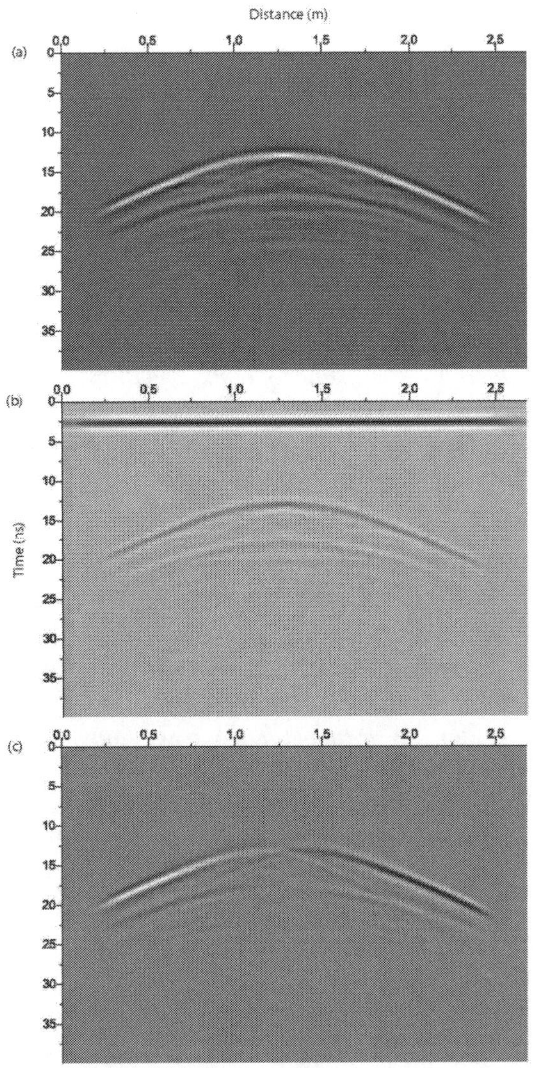

Fig. 6. Synthetic radargrams (as in Fig. 4), corresponding to a void instead of the pig.

References

1. H. M. Jol, *Ground penetrating radar: theory and applications* (Elsevier, 2009)
2. Ł. Ortyl, J. Karczewski, E. Mazurkiewicz, *Importance of selecting parameters of georadar profiles in karst area*, Proc. of 17th International Multidisciplinary Scientific Geoconference: Science and Technologies in Geology, Exploration and Mining., 173-180 (Albena, Bulgaria, 29 June–5 July 2017)
3. S. Tomecka-Suchoń, Acta Geophys. **60**, 386-398 (2012) doi: 10.2478/s11600-011-0019-z
4. M. de la Vega, A. Osella, E. Lascano, J. M. Carcione, Archaeol. Prospect. **12**, 19-30 (2005) doi: 10.1002/arp.244
5. Z. Pilecki, J. Kłosiński, E. Pilecka, J. Karczewski, J. Ziętek, *Wpływ zawodnienia osuwiska na rozpoznanie granic jego nieciągłości metodami falowymi*. Gospod. Surowcami Min. **24**, 427-444 (2008)

6. J. M. Carcione, Geophysics, **61**, 1664-1677 (1996) doi: 10.1190/1.1444085
7. J. M., Carcione, Geophys. Prospect. **44**, 871-888 (1996) doi: 10.1111/j.1365-2478.1996.tb00178.x
8. J. A. Doolittle, N. F. Bellantoni, Journal of Archeological Science, **37**, 941-949 (2010) doi:10.1016/j.jas.2009.11.027
9. J. J. Schultz, M. E. Collins, A. B. Falsetti, J. Forensic Sci. **51**, 607-616 (2006) doi: 10.1111/j.1556-4029.2006.00129.x
10. J. K. Pringle, J. R. Jervis, J. D. Hansen, G. M. Jones, N. J. Cassidy, J. P. Casella, J. Forensic Sci. **57**, 1467-1486 (2012) doi:10.1111/j.1556-4029.2012.02151.x
11. C. A. Dionne, D. K. Wardlaw, J. J. Schultz, *Delineation and Resolution of Cementary Graves Using Conductivity Meter and Ground-Penetrating Radar*. Technical Briefs in Historical Archeology **5**, 20-30 (2010)
12. S. C. Buck, J. Forensic Sci., **48**, 1-7 (2003) doi: 10.1520/JFS2002165
13. T. Golebiowski, Environment Engineering, **24**, 39-53 (2015) doi: 10.4467/2353737XCT.15.225.4611
14. T. Golebiowski, Acta Geophys. **54**, 413-429 (2006) doi: 10.2478/s11600-006-0027-6
15. H. Marcak, T. Golebiowski, *Computer simulation of hydrocarbon flow and wave field for interpretation GPR measurements in contaminated sites*. Proc. of 18th Annual Meeting of SAGEEP, Symposium on the Application of Geophysics to Engineering and Environmental Problems, 837-846, (Atlanta, Georgia, 3-7 April 2005)
16. D. Goodman, Y. Nishimura, *A ground-radar view of Japanese burial mounds*. Antiquity **67**, 349-354 (1993) doi: 10.1017/S0003598X00045403
17. J. M. Carcione, *Wave Fields in Real Media. Theory and numerical simulation of wave propagation in anisotropic, an elastic, porous and electromagnetic media* (Elsevier, 2014)
18. J. Modroo, G. R. Olhoeft, *Ground penetrating radar location of buried avalanche victims*, Proc. of the 2004 International Snow Science Workshop, (Jackson Hole, Wyoming 19-24 Sep. 2004)
19. J. M. Carcione, Geophysics **75**, 53-56 (2010) doi: 10.1190/1.3509472
20. J. M. Carcione, G. Lenzi, S. Valle, Geophys. Prospect. **47**, 1015-1029 (1999) doi: 10.1046/j.1365-2478.1999.00151.x
21. E. Mazurkiewicz, S. Tomecka-Suchoń, R. Tadeusiewicz, Applied Artificial Intelligence **30**, 844-860 (2016) doi: 10.1080/08839514.2016.1274250
22. R. Tadeusiewicz, Arch. Min. Sci. **60**, 971-984 (2015) doi: 10.1515/amsc-2015-0064

E3S Web of Conferences **24**, 01003 (2017) DOI: 10.1051/e3sconf/20172401003
AG 2017 - 3rd International Conference on Applied Geophysics

Relationship between the focal mechanism of magnitude M_L 3.3 seismic event induced by mining and distribution of peak ground velocity

Józef Dubiński[1], *Krystyna* Stec[1*] and *Grzegorz* Mutke[1]

[1]Central Mining Institute, 40-166 Katowice, Poland

Abstract. The relationship between seismic radiation pattern generated by a strong mining induced seismic event and the distribution of peak ground velocity in the epicenter area has been presented. It was a seismic event with the local magnitude M_L = 3.3 occurred on June 21, 2016 in the Upper Silesian Coal Basin (USCB) in Marcel Mine. Calculated values of the peak ground velocity, taking into account the amplification coefficient, were the basis for the development of the PGV_H^{amp} map. The resulting distribution of PGV_H^{amp} isolines and the measured velocity amplitudes point to significant differences. That fact indicates that some additional factors can impact on the seismic effect observed on the surface. One of them could be a focal mechanism of seismic event. Focal mechanism of the M=3.3 induced seismic tremor, were calculated by the moment tensor inversion method. The tremor was characterized by a normal slip mechanism with 87% shear component. Comparison of seismic pattern for *S*-wave at individual stations allowed confirms a relation between directionality of the seismic radiation pattern for *S*-wave and the recorded peak ground velocities and explain the observed anomaly.

1 Introduction

Determining the intensity of earthquakes and of seismic events induced by mining on the surface is an extremely complex issue since the magnitude of the seismic effect on the surface is determined by many factors, including: seismic energy or seismic magnitude of a tremor, epicentral distance, geological structure of the rock mass on the path of seismic waves, seismogeological parameters of the near surface as well as many others [1, 2, 3, 4, 5, 6]. Therefore, it is difficult to estimate seismic intensity (peak ground velocity, *PGV* and acceleration, *PGA*) only as a function of such seismic parameters as seismic energy and epicentral distance. In addition, although one of the most important reasons for the observed local intensity of mining tremors is the amplification of ground motion by the subsurface soil layers, the inclusion of this phenomenon does not completely resolve the observed variability in the distribution of the above-described parameters of mining tremors. Therefore, studies have been undertaken to clarify the large local discrepancies between measured values and prognostic data based on empirical formulas for Upper

*Corresponding author: kstec@gig.eu

Silesia Coal Basin (USCB) [1]. The solutions obtained in this field are not only cognitive but also practical because, due to the high seismicity induced by mining activity, a more precise characteristic of seismic effects on the surface is an extremely significant issue. It is an important element of both the protection of the environment and the formation of a partnership between mining enterprises and local communities, particularly in terms of the acceptance of mining activities. USCB is one of the high seismic areas in Poland induced by mining activity [7, 8, 9, 10, 11, 12]. In the years 2006-2016 there were 13216 shocks with magnitude above 1.6, including 31 very strong (M_L magnitude = 3.26÷4.11) comparable to weak earthquakes. In addition, USCB is a heavily urbanized area, causing seismic impacts, particularly in the case of high-energy tremors, which is a major problem not only in technical but also in social area. For this reason, research into the improvement of research methods and state of the art has been carried out for many years.

2 Ground motions parameters of strong mining induced seismic event of magnitude M_L = 3.3 recorded on June 21, 2016

Two different values of the peak ground velocity PGV_{Hp} and acceleration PGA_{Hp10} were recorded by seismic surface stations in the epicentral distance for the tremor of magnitude M_L = 3.3. The values of the horizontal parameters PGV_{Hp} and PGA_{Hp10} are summarized in Table 1.

Table 1. The ground motions parameters of strong mining induced seismic event of magnitude M_L = 3.3 recorded on June 21, 2016.

Station	PGV_{Hp}, m/s	PGA_{Hp10}, m/s^2 (frequency band up to 10 Hz)	Epicentral distance, m
S1	0.0159	0.535	722
S2	0.0304	1.095	761
S3	0.0555	1.711	115
S4	0.0205	0.542	1100
S5	0.0015	0.035	4,500
S6	0.0012	0.025	5,193

The maximum values of velocity and acceleration on the recorded seismograms were related to the direct transverse waves SP and SH. Figure 1 shows the vibration velocity seismograms on station S2, located in the 761 m from epicentre. Several decades of monitoring of seismic events in the Upper Silesia prove that the highest amplitudes of PGV and PGA from mining seismic events are recorded in the epicenter zone and are derived from transverse waves. For these mining induced seismic event the short duration of the main phase of the vibration is also characteristic (Figure 1). The maximum values of PGV_{Hp} and PGA_{Hp10} were recorded at S3 station at a distance of 115 m. This station was located in a epicentral zone, which explains very high registered values. At the next two stations S1 and S2 located at the distance of 722 and 761m, the velocity amplitudes were smaller but their values differences were two times each other. The subject of research is therefore to determine the cause of this discrepancy, which is not a factor of seismic energy or epicentral distance.

DOI: 10.1051/e3sconf/20172401003

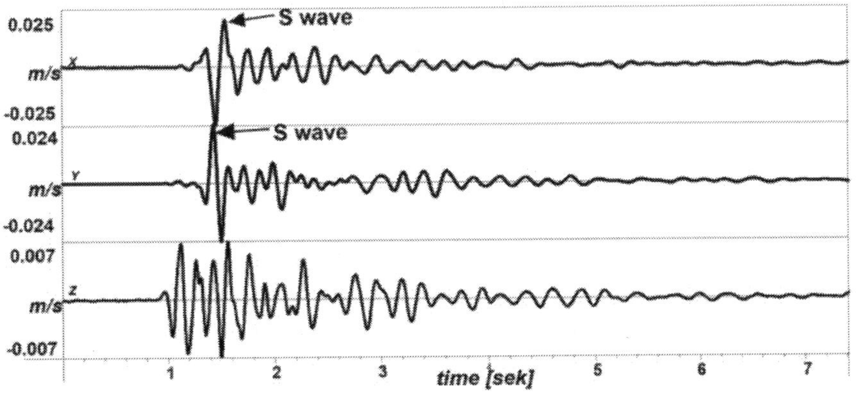

Fig. 1. Velocity ground motion recorded on station S2 in epicentral zone from the mining induced seismic event of magnitude $M_L = 3.3$ on June 21, 2016.

One of the factors influencing the value of ground motions parameters is the geological structure of the near surface quaternary layers, usually characterized by low velocity propagation of seismic transverse wave in Upper Silesia geological condition. The parameter that expresses the influence of quaternary overburden layers on vibration intensity is the W_f amplification factor, which varies in a given location depending on the frequency range and the seismogeological parameters of the overlay layers [1, 3, 7, 13]. In Marcel mine, there is a locally variable quaternary layer (several to tens of meters), characterized by low velocities of seismic waves propagation. Amplification factor W_f was determined based on recognition of lithological of quaternary and tertiary structures and their thickness. In the area of the location of seismic stations there the tremor on June 21, 2016 was recorded, the amplification factor varies from 1.3 to 2.0 (Figure 2). This does not explain, however, the large discrepancies in the values of recorded ground motions parameters at stations S1 and S2.

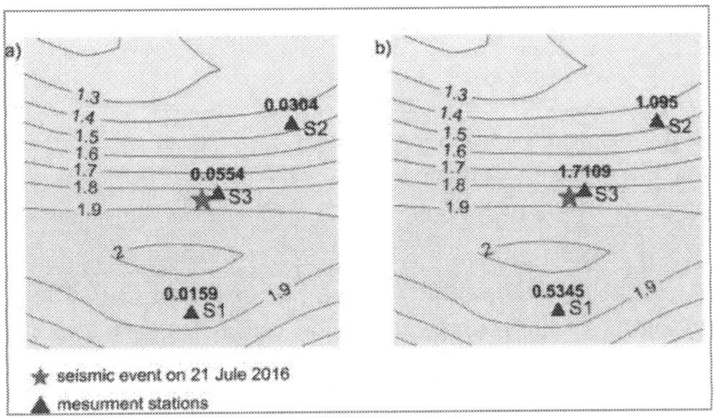

Fig. 2. Distribution of W_f amplification factor and measured values of PGV_{Hp}, m/s (a) and PGA_{Hp10}, m/s^2 (b) for the tremor of $M_L = 3.3$ on June 21, 2016.

3 Distribution of peak ground velocity PGV_{Hmax} for the tremor on June 21, 2016

The assessment of the ground motions intensities generated by induced mining tremors is so far limited to the parameter of amplitude of horizontal vibration velocity PGV_H and time duration. The Mining Seismic Instrumental Intensity Scale MSIIS-15 used in coal mines to study impact effect in buildings are based on these two parameters [12, 14]. In this article we used the method of determining the distribution of PGV_H parameter consists of two basic stages [1].

Determination of the amplitude of horizontal vibration velocity V_{MD} for bedrock foundation in USCB as a function of seismic energy and epicentral distance, for seismic energy E in the range $2 \cdot 10^5 \div 5 \cdot 10^8$ J (magnitude M_L in the range $1.8 \div 3.6$; the energy of the tremors in Polish mines is converted into the local magnitude according to the formula: M = (logE-1,8)/1.9):

$$V_{MD} = [1.48\ 10^{-3}\ (\log E)^{\ 1.23} - 0.011] \cdot [1.55R^{\ 0.135}\exp(-\ 0.77R) + 0.040] \qquad (1)$$

where: E – tremor seismic energy, J,
V_{MD} – maximum horizontal vibration velocity amplitudes of bedrock, m/s,
$R^2 = D^2 + h^2$, D – epicentral distance, km; h – depth of source, km.

Calculation of the horizontal vibration velocity $PGV_H{}^{amp}$ as the multiplication of V_{MD} by the amplification factor W_f:

$$PGV_H{}^{amp} = V_{MD} \cdot W_f \qquad (2)$$

where: W_f – amplification factor for S waves dimensionless [15].

The distribution of velocity isolines according to relation 2 for the analysing seismic event of magnitude M=3.3 is shown in Fig. 3a. As the measured values at individual seismic stations in the epicentral zone deviated from the values determined in accordance with formula 2, the new calculation of the peak ground velocity distribution considers the velocity PGV_{Hp} measured on the surface seismic stations were made. The methodological approach allows for the inclusion of empirical PGV values recorded on six seismic stations. The calculation procedure was as follows [15].

- For each seismic station, an area with a radius $r = 250$ m is determined, within which the $PGV_{Hp}(S_i\, 0)$ value recorded by the seismic devices is assumed.
- In nodal points located at a distance of r>250 m from these stations, the value of corrected amplitude $PGV_H{}^{kor}(r_i)$ of the horizontal velocity is approximated by the formula:

$$PGV_H{}^{kor}(r)_i = d \cdot PGV_{Hp}\ (S_i\, 0) + (1\text{-}d) \cdot PGV_H{}^{amp} \qquad (3)$$

where: $PGV_{Hp}(S_i,0)$ – the value of the peak horizontal vibration velocity measured at the nearest S_i seismic station from a given node, located at r_i from the node, $PGV_H{}^{amp}$ – value calculated from relationship 2 in the node located at distance of r_i from S_i seismic station $d = 250\ / r_i$ for $r_i > 250$m; $d = 1$ for $r_i \leq 250$m.

The results of the $PGV_H{}^{kor}(r_i)$ include the amplitude values recorded at the all seismic stations. The results of the calculations of the corrected peak ground velocity $PGV_H{}^{kor}(r_i)$ are shown in Figure 3b. When analysing this distribution, it is important to note that the distribution of velocity isolines as compared to Figure 3a, as well as the measurements values of the velocity by particular seismic stations are distinctly more compatible. It can be stated that the obtained map of the corrected distribution of the horizontal component amplitude of vibration velocity $PGV_H{}^{kor}(r_i)$ significantly better reflects real state and allows for a more reliable assessment of vibration intensity and its impact on building objects using existing instrumental scales MSIIS-15, which is based on the vibration velocity [12, 14].

DOI: 10.1051/e3sconf/20172401003

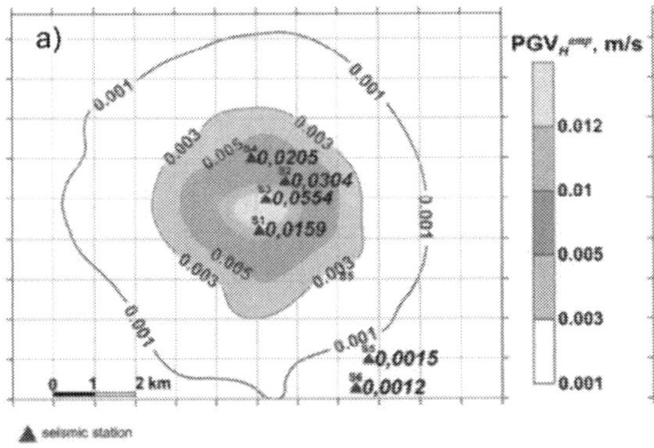

Fig. 3a. Distribution of the velocity amplitudes PGV_H^{amp} calculated according to formula 2 for the mining induced seismic event of magnitude $M_L = 3.3$.

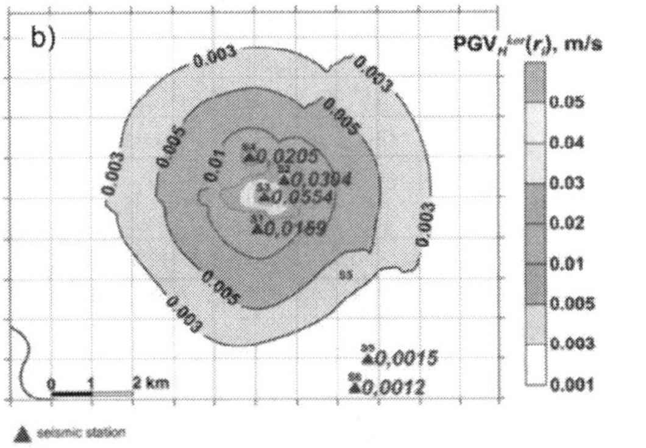

Fig. 3b. Distribution amplitude of the corrected velocity $PGV_H^{kor}(r_i)$ for the mining induced seismic event of magnitude $M_L = 3.3$.

4 Focal mechanism of $M_L = 3.3$ strong mining induced seismic event recorded on June 21, 2016

Excluding the influence of, epicentral distances and amplification factor, it was assumed that the radiations pattern for waves, which depends on the focal mechanism, can be a cause of the discrepancy on PGV_{Hp} and PGA_{Hp10} recorded at measurement station S1 and station S2. The focal mechanism was calculated using the seismic moment tensor inversion method (SMT) in the time domain from amplitudes and polarity of *P*-wave with FOCI software [16]. The SMT analysis was based on the seismograms recorded by underground seismic network of Marcel mine. The 16 stations are horizontally and vertically spaced from the seismic events at distances ranging from 0,5 to 4 km and from 0,1 to 0,8 km, respectively. The stations surrounding the area of the longwall, which was the basis for the correct determination of tremor focus localisation and its mechanism. Angular parameters of the focal mechanism are calculated with error 15°. Figure 4 shows the parameters of the focal mechanism. The full, deviatoric, and pure shear moment tensor was calculated using the L2 norm as a measure of the misfit. This tremor was characterized by a normal slip mechanism with 87% share of the shear component. The remaining explosive component and uniaxial compression were 12 and 1% respectively. As the rupture plane, an azimuth plane was selected located almost at the NE –SW(strike 28°) and a dip of 88° facing to the east. The azimuth of the selected nodal plane could be correlated with the azimuth of the fault zone in this region (Fig. 4).

5 Seismic wave radiation pattern for *S*-wave

Seismic waves radiation pattern in the rock mass depends on the rupture process in the tremor's source. Based on recorded seismograms of tremors, the displacement field in different directions and at different distances from the source area can be determined. A theoretical description of the seismic radiation pattern can be found in Aki [17].

DOI: 10.1051/e3sconf/20172401003

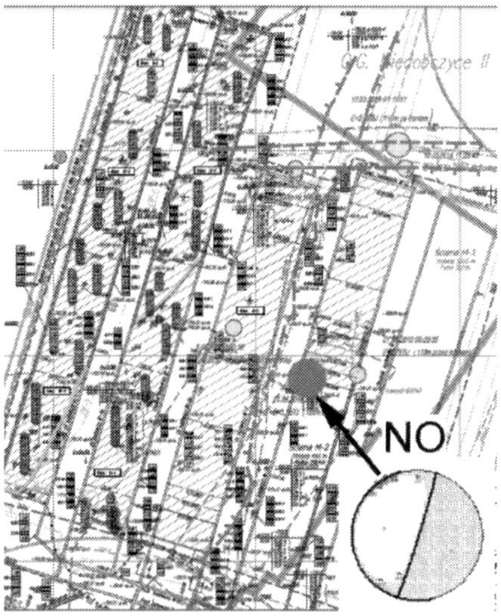

Focal mechanism parameters*

Nodal plane A,B ΦA/δA/λA ΦB/δB/λBA	Axis P,T ΦP/δP ΦT/δT	component TMS, %		
		ISO	CLVD	DBCP
28/88/-89 166/3/-122	289/47 106/43	12	1	87

* ΦA,B – strike A,B; δA,B – dip A,B;
λ – slip angle; ΦP,T – strike P,T;
δP,T – plunge P,T; ISO – explosion; CLVD – compensated linear vector dipole (tension);
DC – double-couple;
NO – normal fault

Fig. 4. Location and moment tensor results for the tremor of $M_L = 3.3$ on the mining map.

Rupture plane dip 0°

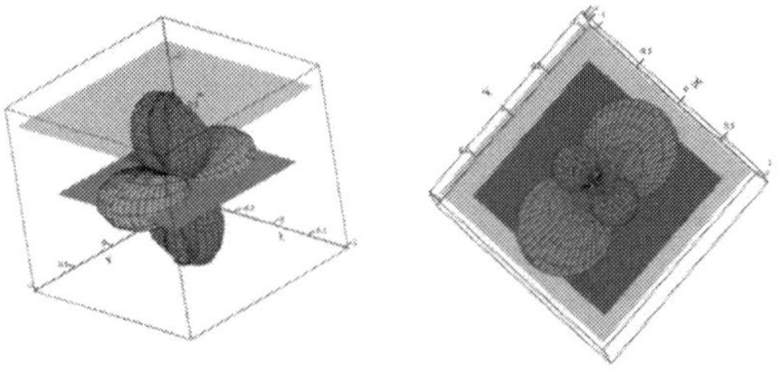

Rupture plane dip 45°

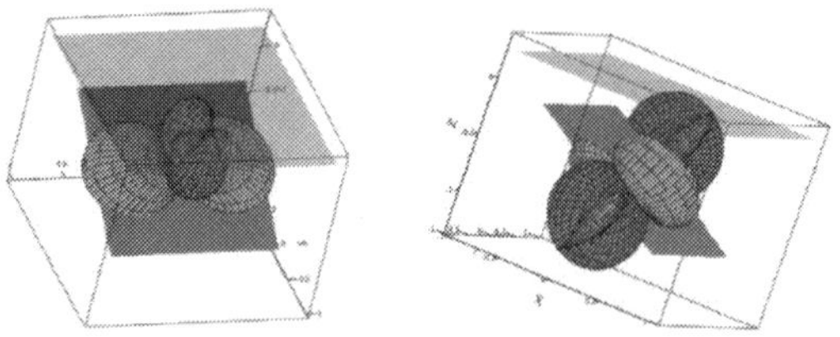

Rupture plane dip 90°

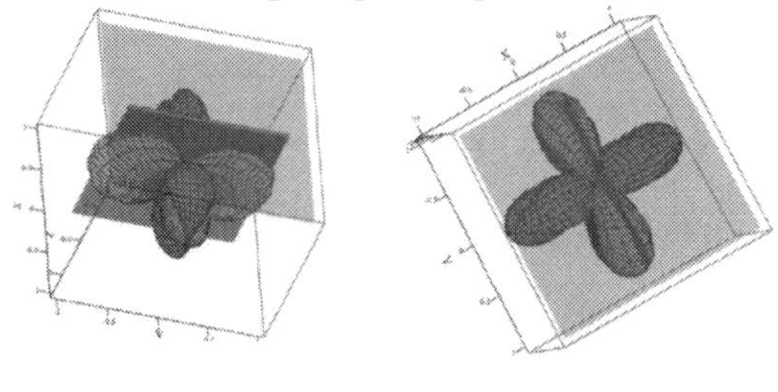

Fig. 5. Visualization of radiation pattern for *SV*- wave (blue) and *SH*-wave (yellow) for the rupture plane with dip of 0°, 45° and 90° (red) with respect to the horizontal plane (green) [20].

Long-term studies of seismic activity in USCB have shown [18, 19] that high energy tremors, the most dangerous for mining excavations, are characterized by shear slip, mathematically described by a double couple of forces. Radiations pattern for seismic waves propagation for the shearing mechanism can be graphically depicted [20]. With this type of focal mechanism are *S*-wave dominate. Figure 5 shows the spatial radiation pattern for the rupture planes with dip of 0°,45°and 90°with respect to the horizontal plane. *S*-wave radiation amplitudes image (waves with the highest amplitudes in the epicentral zone) was showing based on the parameters of the nodal plane A (strike 28°, dip 88°) of the tremor on June 21, 2016.This image along with the registered peak ground velocity values is shown in Figure 6a. The figure shows that the radiation pattern for *S*-wave emitted from the source of the tremor is not uniform, but there are privileged directions which confirm the registrations by more than half of the amplitude of ground vibrations at the S2 station compared to the ground vibrations recorded at the S1 station.

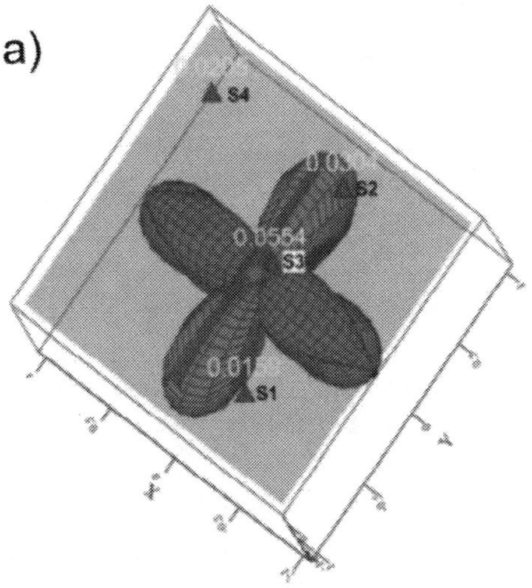

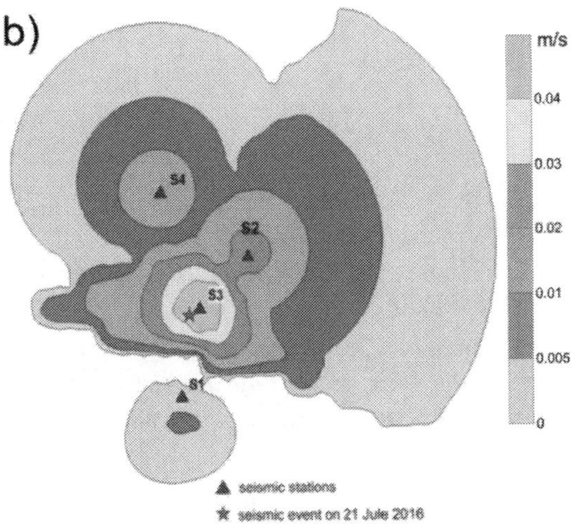

Fig. 6a. Radiation pattern for *S*-wave for the tremor registered on June 21, 2016 of magnitude $M_L = 3.3$ for the focal mechanism shown in Figure 4.

Fig. 6b. Distribution of differential velocity PGV_H^{dif} for tremor on June 21, 2016 of magnitude $M_L = 3.3$.

The theoretical distribution of the seismic radiation pattern for *S*-wave was compared with the "differential velocity amplitudes" PGV_H^{dif} for the tremor registered on June 21, 2016. The image of the "differential velocity amplitudes" PGV_H^{dif} is shown in Figure 6b. This image was obtained by subtracting the distribution of the vibration velocity amplitudes PGV_H^{amp} shown in Figure 3a from the distribution of the vibration velocity amplitudes $PGV_H^{kor}(r_i)$ presented in Figure 3b. As presented, the distribution of "differential velocity amplitudes" PGV_H^{dif} in the epicentral zone is uneven and oriented according to the theoretical distribution radiation pattern for *S*-wave shown in Figure 6b. These calculations have shown that the highly probable factor affecting on the variation in measurement data (PGV_{Hp}, PGA_{Hp10}) at stations located almost at the same distance from the source area of the tremor may be the radiation pattern for *S*-wave, depending on the location of the rupture plane in the seismic event focus.

6 Conclusions

Analyzed case study of strong mining induced seismic event confirms that focal mechanism determining by a directional characteristics of the seismic radiation pattern has also impact on the distribution of horizontal peak ground velocity of vibration PGV_H on the surface.

The obtained results point out that in some cases of strong seismic events focal mechanism can distinctly change distribution of seismic parameters on the surface and consequently intensity of seismic influence.

The paper points out the necessity of elaboration a new empirical relationship with account of focal mechanism of the tremor and presents the developed methodology in this area.

It should be emphasized that solutions obtained in this area are not only cognitive but also practical, mainly due to the possibility of a more precise characteristic of mining activity impacting on environment.

References

1. G. Mutke, J. Dworak, *Factors determining the effect of mining seismic events on buildings in the Upper Silesian Coal Basin. Selected issues of the geophysical studies in the mines – Lubiatów 1991*, Publs. Inst. Geophys. Pol. Acad. Sc. **M-16**, 245, 115-130 (in Polish) (1992)
2. Z. Zembaty A. Rutenberg, Eng. Struct. **24**, 11, 1485-1496 (2002) doi: 10.1016/S0141-0296(02)00096-2
3. D. Olszewska, S. Lasocki. *Application of the horizontal to vertical spectral ratio technique for estimating the site characteristics of ground motion caused by mining induced seismic events.* Acta Geophysica Polonica, **52**, 3, 301-318 (2004)
4. J.G. Anderson, BSSA. **100**, 1, 1-36. doi: 10.1785/0120080375 (2010)
5. S. Lasocki, Acta Geophysica Polonica **61**, 5, 1130-1155 (2013) doi:10.2478/s11600-013-0139-8
6. J. Chodacki, Acta Geophysica Polonica **64**, 6, 2449-2470 (2016) doi: 10.1515/acgeo-2016-0109
7. G. Mutke, K. Stec, *Seismicity in the Upper Silesian Coal Basin, Poland: Strong regional seismic events.* Proc. 4th Int. Symp. - Rockbursts and Seismicity in Mines, 213-217 (Rotterdam 1997)
8. A.F. Idziak, L. Teper, W.M. Zuberek, *Seismic activity and tectonics of the Upper Silesian Coal Basin Katowice*. Publ. of Siles. Univ. (in Polish) (1999)
9. H. Marcak, G. Mutke, J. Seismol. **17**, 4, 1139-1148, (2013) doi: 10.1007/s10950-013-9382-3
10. M. Kozłowska, B. Orlecka-Sikora, Ł. Rudziński, Sz. Cielesta G. Mutke, Int. J. Rock Mech. Min. **86**, 5-15 (2016) doi: 10.1007/s00024-016-1432-7
11. K. Stec, Geophys. J. Int. **168,** 2, 757-768 (2007) doi: 10.1111/j.1365-246X.2006.03227.x
12. G. Mutke, J. Dubiński. *Seismic intensity induced by mining in relations to weak earthquakes.* Proc. of the 24th World Mining Congress. Part. Underground Mining, 399-407 (Rio de Janeiro 2016)
13. E.F. Savarienskij, *Evaluation of the impact of near surface layer on the amplitude of ground motion*, Izv. Akad. Nauk SSSR Geofizyka **10**, 1441-1447 (in Russian) (1959)

14. G. Mutke, J. Chodacki, L. Muszynski, S. Kremers, R Fritschen, *Mining Seismic Instrumental Intensity Scale MSIIS-15 – verification in coal basins*, AIMS 2015 - Fifth Int. Symp. Mineral Resources and Mine Development **14**, 551-560 (RWTH Aachen University 2015)
15. G. Mutke, H. Marcak, F. Mutke, A. Barański, *The method of determining the distribution maps of seismic intensity IGSI after recording of a strong seismic event induced by mining,* Przegląd Górniczy, **2**. 51-58 (in Polish) (2017)
16. G. Kwiatek, P. Martínez-Garzón, M. Bohnhoff, Seismol. Res. Lett., **87**, 4 (2016) doi:10.1785/0220150251
17. K. Aki, P.G. Richards, *Quantitative Seismology – Theory and Methods* (W. H. Frejman & Co., San Francisco 1980)
18. K. Stec, Arch. Min. Sci. **57** 4, 871–886 (2012) doi: 10.2478/v10267-012-0057-7
19. K. Stec, Journal of Sustainable Mining, **1/2015**, 55-65 (2015) doi: 10.1016/j.jsm.2015.08.008
20. A. Lurka, *Kierunkowość radiacji sejsmicznej (Directionality of the seismic radiation pattern)* in: *Geologiczne przyczyny wzmacniania drgań w nadkładzie serii węglowej na obszarze Górnośląskiego Zagłębia Węglowego* (ed. K. Stec), 43-46 (in Polish) (CMI, Katowice, 2013)

E3S Web of Conferences **24**, 01004 (2017) DOI: 10.1051/e3sconf/20172401004
AG 2017 - 3rd International Conference on Applied Geophysics

The influence of regional geological settings on the seismic hazard level in copper mines in the Legnica-Głogów Copper Belt Area (Poland)

Zbigniew Burtan[1*]

[1]AGH University of Science and Technology, 30-059 Krakow, Poland

Abstract. The current level of rockburst hazard in copper mines of the (LGOM) Legnica- Głogów Copper Belt Area is mostly the consequence of mining-induced seismicity, whilst the majority of rockbursting events registered to date were caused by high-energy tremors. The analysis of seismic readings in recent years reveals that the highest seismic activity among the copper mines in the LGOM is registered in the mine Rudna. This study investigates the seismic activity in the rock strata in the Rudna mine fields over the years 2006-2015. Of particular interest are the key seismicity parameters: the number of registered seismic events, the total energy emissions, the energy index. It appears that varied seismic activity in the area may be the function of several variables: effective mining thickness, the thickness of burst-prone strata and tectonic intensity. The results support and corroborate the view that principal factors influencing the actual seismic hazard level are regional geological conditions in the copper mines within the Legnica-Głogów Copper Belt Area.

1 Introduction

Major natural hazards in copper mining in Poland include rock bursts and tectonic movements [1-2]. Most rock bursts and rock mass destressing events registered in the (LGOM) Legnica-Głogów Copper Belt Area have been the consequence of high-energy events whilst the current rock bursting hazard level is associated with mining induced seismicity. Potential of high-energy tremors is an inherent feature of copper fields within the LGOM area mostly due to the occurrence of burst-prone roof rock strata over the worked-out fields and periodic faulting movements, these impacts are further exacerbated due to current mining activities [2-5].

The analysis of seismic readings registered in recent years in the LGOM area suggests that Rudna is the most tectonically disturbed of the mines in the area [6], which is the result of natural conditions (mining depth, thick-layered rock mass structure, faulting) and the need to mine plots in areas already disturbed by extensive mining operations.

* Corresponding author: burtan@agh.edu.pl

2 Local settings of the Rudna mine

The Rudna mine is one of the three mines of the Polish Copper Corporation KGHM Polska Miedź S.A. located within the Legnica-Głogów Copper Belt Area, to the north of the town Polkowice, its mineable horizon being 77.8 km2 (Fig 1). The average production capacity now is about 12 mln tons of copper per year [7].

Fig. 1. Copper mines within the LGOM area in Poland [7].

The Legnica-Głogów Copper Belt Area is a part of the Fore-Sudetic Monocline and its geological structure and tectonic features are determined by particular stages of sedimentation processes and rock mass dislocations [8]. The copper ore deposit there is tectonically disturbed, experiencing seismic activity. There are some major faults with a throw of 100 m as well as associated groups of rolls and smaller faults that have the throw of only about 10 m. Of particular importance are the faulting zones in the Rudna Główna site (Central Rudna) and in Biedrzychowa [7].

Copper ore deposits include three rock types with different lithology: red footwall sandstone, illite-dolomite shales and zechstein dolomites. The deposit thickness within the Rudna mine varies from 0.5 to 20 m (4.3 m on the average), 20% of these being thick beds with thickness exceeding 10 m. Roof rock strata include dolomites and limestone 12.2-

106.3 m in thickness, their compression strength up to 25 m over the roof varies from 30 to 250 MPa (113 MPa on the average) [7].

Mining operations are continued in three mine areas: Central Rudna (RG), Western Rudna (RZ) and Northern Rudna site (RP). There are 11 shafts accessing the deposit and mining operations are carried out at the depth 850-1250 m [7].

Copper deposit in Rudna is mined by the room and pillar method with the roof control strategy for deposits up to 7 m in thickness, whilst hydraulic backfill system is used to facilitate the mining of thicker ore zones (more than 7 m). The blasting technology is used during the development and preparatory works as well as in the course of regular mining activities [7].

3 Seismicity

The historical frequency of seismic activity in the mine field areas of the Rudna mine site is investigated, covering the years 2006-2015 [6]. The level of seismic activity, measured by the number of tremors N in particular energy categories (A_s>10^3 J), the total number of tremors ΣN, the total seismic energy ΣA_s and the energy expense (in relation to the output level) J, is shown in Table 1. Total seismic energy and the energy expense in the analysed mine regions are illustrated in Fig 2 and Fig 3.

Table 1. Seismicity in the proximity of the Rudna mine over the years 2006-2015.

Mining area	Number of tremors N [-]							ΣN	ΣA_s	J
	10^3	10^4	10^5	10^6	10^7	10^8	10^9	[-]	[GJ]	[J/Mg]
Central Rudna	6188	2294	816	283	57	5	1	9644	5.7	137.5
Western Rudna	5272	2236	738	194	54	2	-	8496	2.6	72.5
Northern Rudna	3363	1509	492	157	55	3	-	5579	2.5	76.8

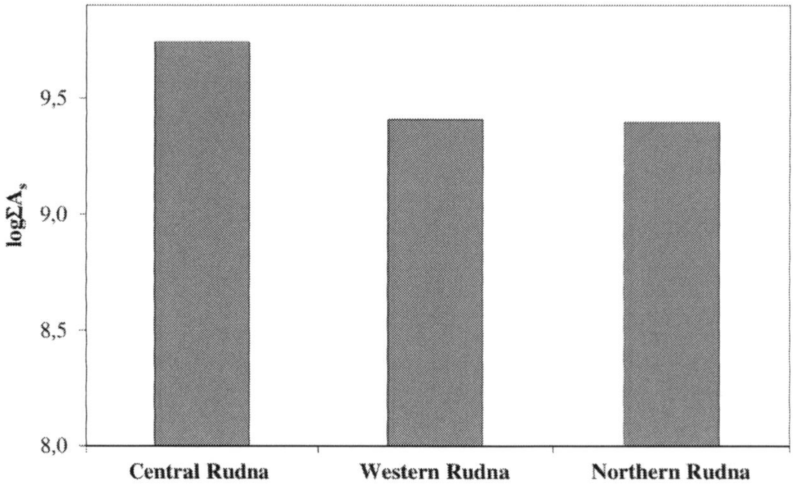

Fig. 2. Total seismic energy within the Rudna mine over the years 2006-2015.

DOI: 10.1051/e3sconf/20172401004

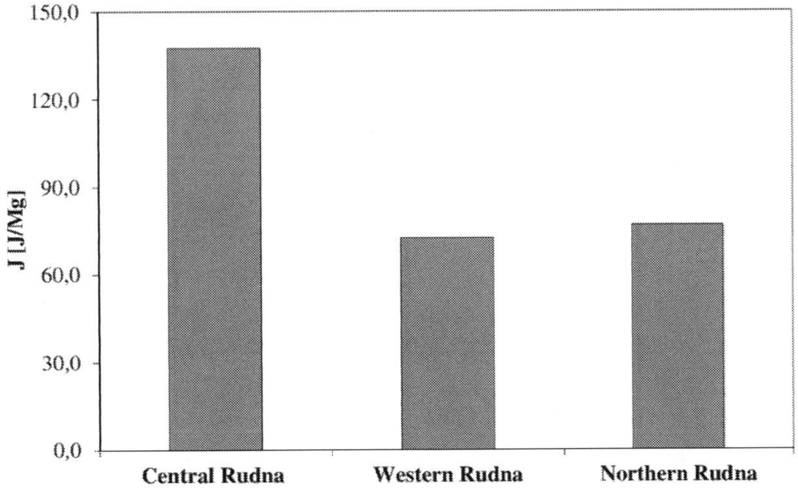

Fig. 3. Energy expense indicator in the Rudna mine site over the years 2006-2015.

It appears that the seismic hazard level in particular mines sections varied over the years 2006-2015. No matter which seismicity parameter was considered, the highest level of seismic hazard would be registered in the Central Rudna site whilst the hazard levels in the Western Rudna and Northern Rudna sites were lower (both in quantitative terms and in terms of energy expense). In terms of unit energy expense, the ordering of the two latter sites would be reversed (Northern Rudna, Western Rudna).

4 Influence of geological settings

Questing the reasons for varied seismic hazard level in the Rudna site, reference is made to geological settings in the context of the mining thickness associated with the ore deposit thickness, actual thickness of rock strata in the proximity of the mining operations and tectonic intensity.

4.1 Ore deposit thickness

A principal factor influencing the seismic hazard level is the ore deposit thickness and associated height of cut. A body of data gathered in the LGOM copper mines and in research reports indicates that increasing the cutting height and removing the old excavations by the same strategy results in enhanced seismic hazard levels [2,4,9].

The study of representative geological profiles (Fig 4) supported by a brief summary of mining activities to date (Table 2) yields the effective mining thickness in particular sites within the Rudna mine [7]:

- from 5.1 to 16.5 m (11.9 m on the average) in Central Rudna;
- from 3.7 to 14.6 m (5.5 m on the average) in Western Rudna;
- from 2.8 to 18.3 m (8.4 m on the average) in Northern Rudna.

These findings are in agreement with the effective mining thickness distribution in the worked-out field section in the Rudna mine (Fig 5) [6].

Recalling the previously derived dependence and the seismic readings data, it is reasonable to suppose that greater ore deposit thickness and height of the cut in the Central Rudna region may be responsible for the highest seismic hazard level compared to other mine sections.

 DOI: 10.1051/e3sconf/20172401004

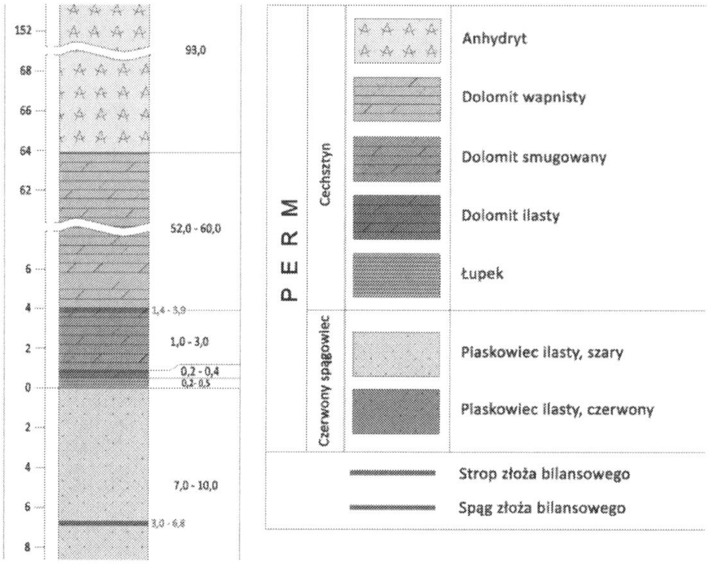

Fig. 4. A geological profile of the deposit in the Rudna mine [6].

Table 2. Effective mining thickness in various section of the Rudna mine.

Mining area	Section	Mine field	Mining thickness [m]	Ripped dolomite strata [m]	Residual dolomite [m]
Central Rudna	G-1	XII/1	14.1	2.5	87.5
		XII/2	12.4	0.6	86.9
	G-2	XVII/1	12.9	0.8	71.1
		XVII/2	8.3	1.5	72
		XI/1	10.7	3.4	60
		G-15/10	5.1	1.8	40
	G-3	G-3/4	13.6	2.9	77
		G-1/7 Blok A	16.5	1.2	95
	G-7	G-7/5	16.1	2	70
		XIV/2	9	2	85
Western Rudna	G-11	XIX/1	5.7	1.6	17.7
	G-12	XIX/2	4	1.4	17.9
	G-14	XXV/2	4.4	1.2	9.8
	G-15	XVI/1	14.6	1.4	57.6
		XV/2	4	0.8	57.2
		XXV/1	3.7	1	14
	G-17	XV/1	4.9	2.3	18.6
		XV/4	4.5	1.9	12.5
		XV/5	3.7	0.8	10
Northern Rudna	G-23	XXI/1	5.6	0	54.4
		XXI/2	11.7	0.2	67
	G-24	XXIII/2	4.6	1.4	18.7
	G-25	XXVIII/1	2.8	0.3	15.7
	G-26	XXIII/3	18.3	3.7	45
		XXIII/4	7.4	2.7	30

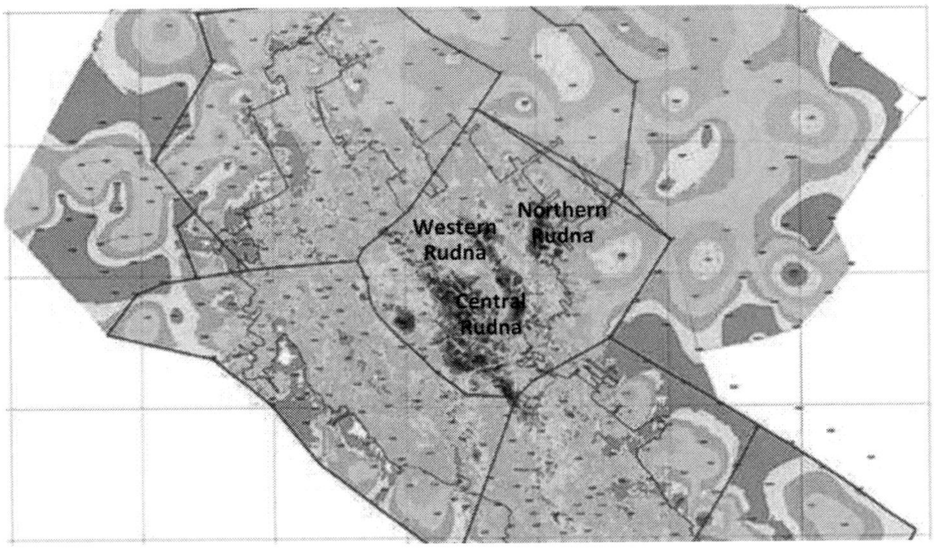

Fig. 5. Cutting height distribution in the Rudna mine [6].

4.2 Thickness of burst-prone strata

Another major determinant of the seismic hazard level is the occurrence of burst-prone high-strength roof rock strata with considerable thickness above the mining horizon. As a result of mining operations, these strata tend to bend and crack. The presence of such strata and associated occurrence of rock burst events depends on the thickness of this roof strata which, in turn, determines the amount of seismic energy generated during the cracking. The thicker the damaged strata, the greater seismic energy would be released during the tremor [9]. Burst-prone strata in the Rudna mine are dolomite strata with the compression strength of about 100 MPa. The thickness of dolomite strata varies, ranging from 12.2 m in the north-west part of the ore deposit (Western Rudna) to 106.3 m in the western and central part of the ore deposit (Central Rudna); its thickness rapidly decreases from 40-60 m to less than 20, in the north-west part of the mine field [6,7]. The summary of excavated and abandoned dolomite rock strata (Fig 4) in recently mined fields in the Rudna mine (Table 2) indicates that their thickness falls in the range [6]:

- from 40 to 95 m (74.5 m on the average) in Central Rudna;
- from 9.8 to 57.6 m (23.9 m on the average) in Western Rudna;
- from 15.7 to 67 m (38.5 m on the average) in Northern Rudna.

It is considered likely that smaller thickness of dolomite rock strata in this area is influential in causing lower seismic activity in the Western Rudna and Northern Rudna sites.

4.3 Tectonic movements

The occurrence of faulting zones in the Rudna mine, which can be displaced during the mining operations thus generating high-energy shocks, in the major factor determining the seismic hazard level. Seismic data on the mine sections abutting the faulting zones of the Central Rudna (striking north-west – south east) and in Biedrzychowa (striking NNE-SSW) (Fig 6) indicate that the seismic hazard level measured by the proportion of high-energy events in the faulting zones was higher than in the remaining sections of the Rudna mine [4]. One has to bear in mind that the fault zones occur in the active mining areas of the Rudna mine (Fig 6), so the high level of seismic hazard in the area is also the consequence of intensive tectonics.

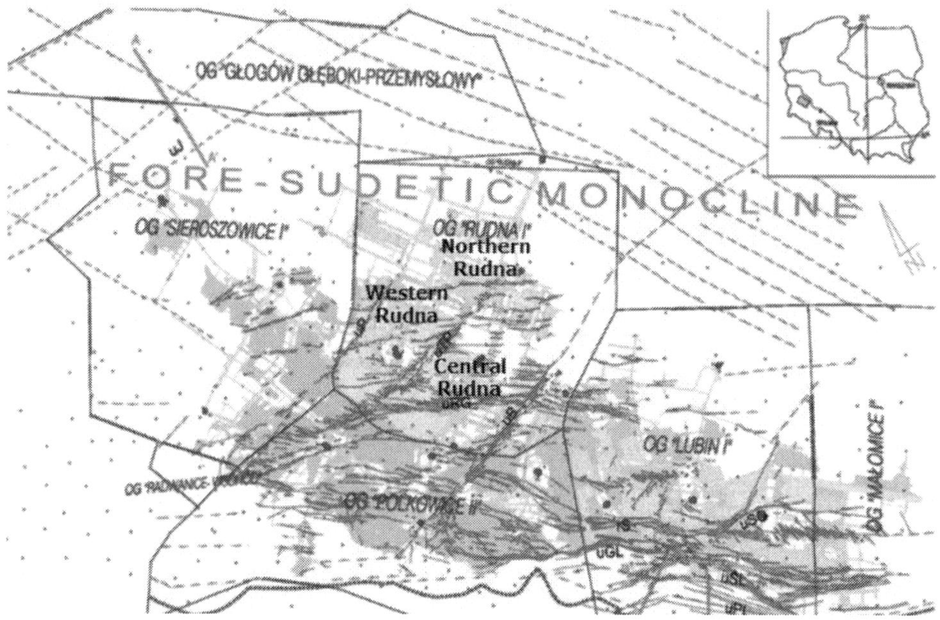

Fig. 6. Fault zones in the Rudna mine [4].

5 Conclusions

The study of seismicity in the Rudna mine over the period 2006-2015 leads us to the following conclusions: –

- The seismic hazard in the entire mine, measured by the number of registered seismic events, the total seismic energy released by the rock strata and the unit energy expense, remained on a high level, comparable in recent years.
- The level of seismicity of rock strata in active mining areas within the mine was varied. The highest seismicity was registered in the Central Rudna section, seismicity levels in Western Rudna and Northern Rudna were lower, which can be attributed to the differences in effective mining thickness and in thickness of burst–prone roof rock strata as well as varied tectonic intensity in those areas.

This study confirms the view that the principal factors influencing the seismic hazard level in the LGOM copper mines are regional geological settings. It is worthwhile to mention, therefore, that the high level of seismic and rock burst hazard in LGOM copper mines, including Rudna, is the consequence of the mining depth as well as most difficult mining conditions: intensive mining operations, mining the residues as well as support and protecting pillars and ore mining in the areas abutting old excavations.

References

1. J. Dubiński, Z. Pilecki, W. Zuberek, eds., *Badania geofizyczne w kopalniach – przeszłość, teraźniejszość i zamierzenia na przyszłość* (IGSMiE PAN, Kraków, 2001)
2. A. Goszcz, *Wybrane problemy zagrożenia sejsmicznego i zagrożenia tąpaniami kopalniach podziemnych* (Biblioteka Szkoły Podziemnej Eksploatacji Złóż, Wydawnictwo Nauka Technika, Kraków 2004)
3. J. Butra, *Eksploatacja złoża rud miedzi w warunkach zagrożenia tąpaniami i zawałami* (KGHM Cuprum Sp. z.o.o. CBR, Wrocław 2010)

4. Z. Burtan, *Wpływ eksploatacji w rejonach zaburzeń uskokowych o dużych zrzutach na kształtowanie się zagrożenia sejsmicznego w kopalniach Legnicko Głogowskiego Okręgu Miedziowego,* Rozprawy Monografie **247** (Wydawnictwa AGH, Kraków 2012)
5. E. Pilecka, Z. Pilecki, *Analysis of relation between induced seismic activity and satelite data,* Proc. 19th Symposium on Application of Geophysics to Engineering and Environmental Problems SAGEEP, EEGS, 346-355 (Seattle, USA, 2-6 April 2006)
6. *Materiały dotyczące warunków geologiczno-górniczych i aktywności sejsmicznej górotworu w KGHM Polska Miedź S.A, oddział ZG Rudna* (KGHM Polska Miedź S.A., Oddział ZG Rudna Polkowice, 2006-2015)
7. *Report on the Mining Assets of KGHM Polska Miedź S.A. located within the Legnica-Głogów Copper Belt Area (*KGHM Polska Miedź S.A. Lubin, 2012)
8. J. Ślizowski, Z. Pilecki, K. Urbańczyk, E. Pilecka, L. Lankof, R. Czarny, Adv. High Energy Phys. (2013) doi: 10.1155/2013/461764
9. A. Zorychta, *Geomechaniczne przesłanki doboru metod minimalizacji zagrożenia wstrząsowego,* Bezpieczeństwo Pracy i Ochrona Środowiska w Górnictwie **7** (1997)

An application of the geophysical methods and ALS DTM for the identification of the geological structure in the Kraśnik region – Lublin Upland, Poland

Mirosław Kamiński[1,*]

[1]Polish Geological Institute – National Research Institute, Geological Mapping and Regional Geology Program, 00-975 Warsaw, Poland

Abstract. The purpose of the study was the assessment of the viability of selected geophysical methods and the Airborne Laser Scanning (ALS) for the identification and interpretation of the geological structure. The studied area is covered with a dense forest. For this reason, the ALS numerical terrain model was applied for the analysis of the topography. Three geophysical methods were used: gravimetric, in the form of a semi-detailed gravimetric photograph, Vertical Electrical Sounding (VES), and Electrical Resistivity Tomography (ERT). The numerical terrain model enabled the identification of Jurassic limestone outcrops and interpretation of the directions of the faults network. The geological interpretation of the digitally processed gravimetric data enabled the determination of the spatial orientation of the synclines and anticlines axes and of the course directions of main faults. Vertical Electrical Sounding carried along the section line perpendicular to the Gościeradów anticline axis enabled the interpretation of the lithology of this structure and identification of its complex tectonic structure. The shallow geophysical surveys using the ERT method enabled the estimation of the thickness of Quaternary formations deposited unconformably on the highly eroded Jurassic limestone outcrop. The lithology of Quaternary, Cretaceous and Jurassic rocks was also interpreted.

1 Introduction

Recent years brought an increase in the practical implementation of modern geophysical and photogrammetric methods in geological charting [1]. Geophysical method are used for the identification and interpretation of geological structure while the photogrammetric methods are often used to analyse the topographic formations.

Among the photogrammetric methods, particular attention should be paid to the numeric terrain model (NTM) generated as a result of filtering a cloud of points received from Airborne Laser Scanning. The advantage of this model is detailed representation of the surface topography. The NTMs are generated only with points representing the surface. For this reason, ALS NTMs are particularly useful for the geomorphological surveying of limited

* Corresponding author: miroslaw.kaminski@pgi.gov.pl

access areas and areas covered by vegetation and forests. The accuracy of the model depends on scanning density. For example, the ISOK project (IT system of the Country's Protection Against Extreme Hazards) uses the scanning density for non-urban areas of 4 points/m^2. Further computer analysis of this model enables appropriate NTM visualisation. It is possible to track the readability of lithology (resistance to weathering) and tectonic structure (determination of the positions of strata, course of faults and folds. This model is particularly useful for the interpretation of various geomorphological formations which is very important for geological charting. It enables the range and shape of river terraces, dunes, ravines and mass movements, particularly landslides [1, 2].

Other non-invasive geophysical method are also used in order to prepare the Detailed Geological Map of Poland in the 1:50,000 scale. Some of them are already included in the typical set of surveying methods in geological charting, others are just being introduced. Electrical resistivity sounding and profiling methods are among the more important. The Electrical Resistivity Tomography (ERT) method is less often used for geological charting.

The semi-detailed analysis of a gravimetric photograph is also used for geological surveying. The area of Poland is covered with such a photograph with a density of nearly 3 points per km^2. Each point is provided with gravimetric data, geographical coordinates and elevation data. These data enable the creation of gravimetric and topographic maps. The use of gravimetry in geological charting is mostly based on the analysis of the maps of gravimetric anomalies and their qualitative interpretation. In effect, the information regarding the distribution of geological formations and courses of dislocation zones are obtained [3].

2 Geomorphological and geological characteristics of the study area

The area of the Kraśnik sheet of the DGMP (Detailed Geological Map of Poland) in the scale of 1:50,000 is located in the south-eastern part of Poland, within the Lublin Upland (Fig. 1). According to the physico-geographical division of Poland [4], the area of the sheet is located entirely within the Urzędów Heights. The Urzędów Heights comprise the south-western part of the Lublin Upland. They are an upland with levelled plateaus, cut with river valleys and ravines typical for loess areas. The most prominent feature of the topography is the long Wyżnica river valley with asymmetric slopes. The left slope, consisting of loess formations, is significantly higher and steeper. The right slope, consisting of Cretaceous formations, is more gentle and blends in with the topography. The elevation differences within the sheet area are relatively significant, upwards of 100 m.

In terms of geology, the study area is located within the Radom-Kraśnik Palaeozoic elevation which forms a foundation for the SW wing of the Lublin basin [5]. The basin separates the rigid East-European platform from the fold structures of Western Europe. The Palaeozoic foundation is located at the depth from 1000 to 1500 m. It consists of Devonian limestones and dolomites as well as Carboniferous mudstones, claystones and sandstones. Upper Jurassic formations are deposited unconformably on top of the foundation. Higher in the profile, there is a complex of Upper-Cretaceous carbonate rocks with the thickness of approx. 800–1000 m. Upper Cretaceous formations are represented by Cenomanian, Turonian, Coniacian, Santonian, Campanian and Maastrichtian deposits. In this area, Quaternary deposits are only partially preserved. The north-eastern part of the area is heavily denudated and devoid of these formations in many places. The south-western part is covered with a thick layer of loess (max. over 10 m). Under the loess layer, there are Riss glaciation formations, i.e. glacial sands and gravels, residual sands with boulders and glacial tills.

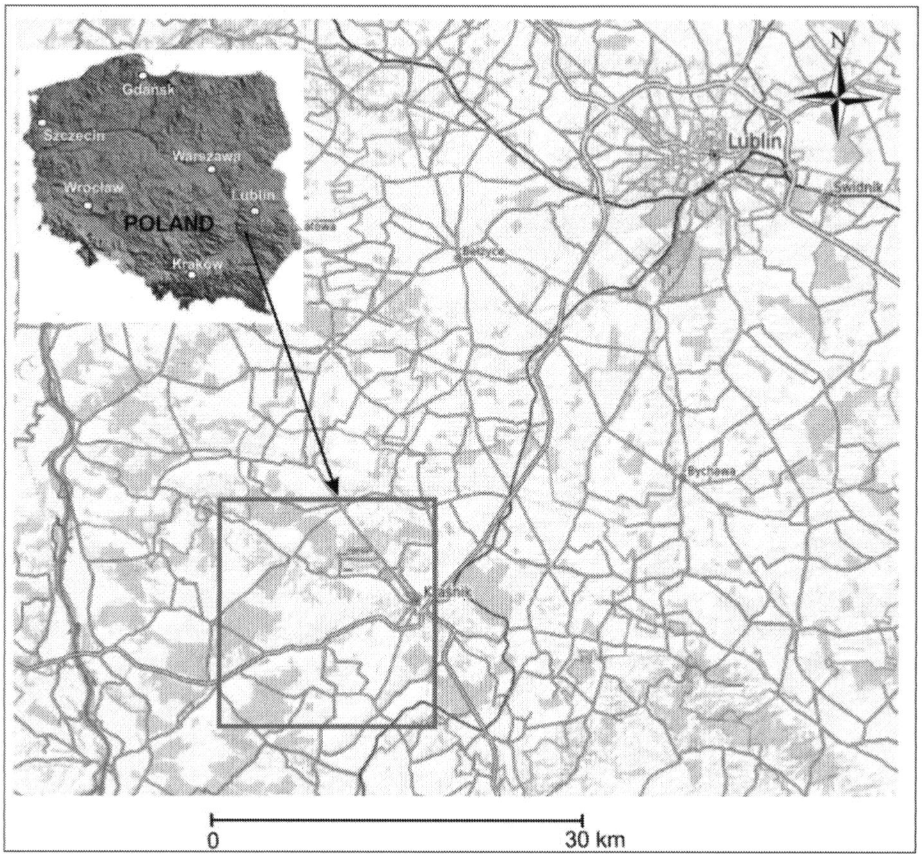

Fig.1. Location of the study area.

3 Methodology

3.1 Gravimetry

The gravimetric method analyses the changes in the intensity of the weight field which, in some simplification, correspond to the density variation of rocks forming the analysed rock medium [3]. The changes in the rock density value are a result of the lithological changes of the rock medium. The use of gravimetry in geological charting is mostly based on the analysis of the maps of gravimetric anomalies and their qualitative interpretation. In effect, the information regarding the distribution of geological formations and courses of dislocation zones are obtained. The primary map used for the analysis is the Bouguer anomaly map. The map of residual anomalies obtained by removing regional anomalies from the Bouguer anomaly map is more useful for the purposes of shallow geological charting. In this method we obtain a map of variation in the gravitational field in the zone from the surface to the depth determined as the function of the assumed averaging radius. For the DGMP Kraśnik sheet, the gravimetric maps were prepared using the Surfer v.11 software (Golden Software Inc.). The interpolation grids with the mesh size of 250 x 250 m were calculated using kriging. The map of anomalies in the Bouguer reduction was calculated for the reduction density of 2.25 g/cm^3. The map of local anomalies was calculated using the Griffin method [6], assuming the averaging radius of r=1km.

 DOI: 10.1051/e3sconf/20172401005

3.2 Vertical Electric Sounding (VES)

Vertical Electric Sounding (VES) involves measuring the electrical resistivity with varying distances of current electrodes from potential electrodes in consecutive measurement steps. As the distance between the electrodes is increased, the depth range of the measurement increases. The resulting chart presents the distribution of apparent electrical resistivity from the ground surface down [7]. The quantitative interpretation of the sounding involves the determination of the thickness and specific resistivity of geological strata [8]. The VES measurements were carried out using the Schlumberger symmetrical measurement system. The length of the power supply line to the measurement system was AB/2=250-320 m, which provided the depth range of measurements of 130–150 m. The measurements were carried out using the direct current digital measurement instruments type ADG-7 powered by batteries with a direct current converter.

Sounding location and terrain morphology were determined based on GPS measurements and the topographic map in the scale of 1:25,000. The results of the measurements were presented on a geoelectrical section made in the horizontal scale of 1:20,000 and vertical scale of 1:2000. The section presents the location and distribution of the RoT resistivity values for each VES measurement profile. For some sections of the study area, the resistivity of near-surface layers reach tens of thousands of ohms. For this reason, it was decided to present the results in the form of logarithmic values, for better graphical representation of the resistivity values distribution. The resulting material of the field studies is the RoT electrical resistivity section presenting the RoT values to the depth of 200 m.

3.3 Electrical Resistivity Tomography (ERT)

The Electrical Resistivity Tomography method provides continuous values of electrical resistivity along the measurement profile line due to the placement of a large number of measurement electrodes [9]. The electrodes are installed at even distances and connected by a multi-wire cable with a computer electrode selector and an electric meter. The electrode selector enables the connection of the meter to any combination of electrodes and measuring the apparent resistivity. Then, further measurements are taken according to the programmed procedure. It is possible to select any measurement system, e.g. Wenner, Dipole-Dipole or Schlumberger. The measured values of apparent resistivity are processed and visualised for purposes of qualitative and quantitative interpretation. The Res2DINV software by Geotomo Software was used to process the data and carry out the inversion.

As part of the field study, an ERT profile was prepared on the Kraśnik sheet in the Wenner measurement system with the length of 600 m and electrodes pitch of 5 m. The maximum depth range was approx. 60 m.

3.4 ALS digital model

The elevation data from the flood-prevention ISOK project were used for studying the topography. The products of the ISOK project are provided by the Geodesic and Cartographic Documentation Center in Warsaw (www.codgik.gov.pl). The data obtained during laser scanning are save in the form of ASPRS LAS files, called LAS (Laser File Format). The scanning density was 4 points per square metre. The primary processes for processing the points cloud provided by the ALS scan are classification and filtration [10]. Elevation data were acquired in September 2014. The NTM generation process involved the classification of the points cloud. The points cloud was divided into the following layers: surface, low vegetation, medium vegetation, high vegetation and buildings. Then, the tools available in the QCoharent software were applied to filter the points related to surface topography from the cloud. The next step was the interpolation of elevation points using the deterministic algorithm IDW (Inverse Distance Weight). The effect was a NTM with the grid mesh size of 0.5 m.

4 Results

4.1 Gravimetry

The analysed area is located within the gravimetric image of the Lesser Poland Upland [11], with visible three smaller positive anomalies. The Kraśnik sheet is located on the border of the anomaly called the Lublin Upland or the Janów-Biłgoraj Anomaly and the local lowland internally dividing the Lesser Poland Upland [12]. The Bouguer anomaly map presents four positive and three negative anomalies. Comparing the results with the structural map of the Cretaceous floor [13], it was found that the positive anomalies correspond to the courses of anticlines while the negative anomalies are correlated with the courses of the synclines. As a result, four anticlines and three synclines are presented on the Kraśnik sheet of the DGMP (fig. 2). The map of residual anomalies was used to interpret the directions of the fault network. The spatial distribution of the faults network (fig. 6B) was interpreted based on the archival geological data [14, 15] and the analysis of density differences in residual anomalies. Three directions were identified. In the southern part of the sheet, the dominant directions are N-S and W-E, while in the northern part, the most common directions are NE-SW and NW-SE (Fig. 3).

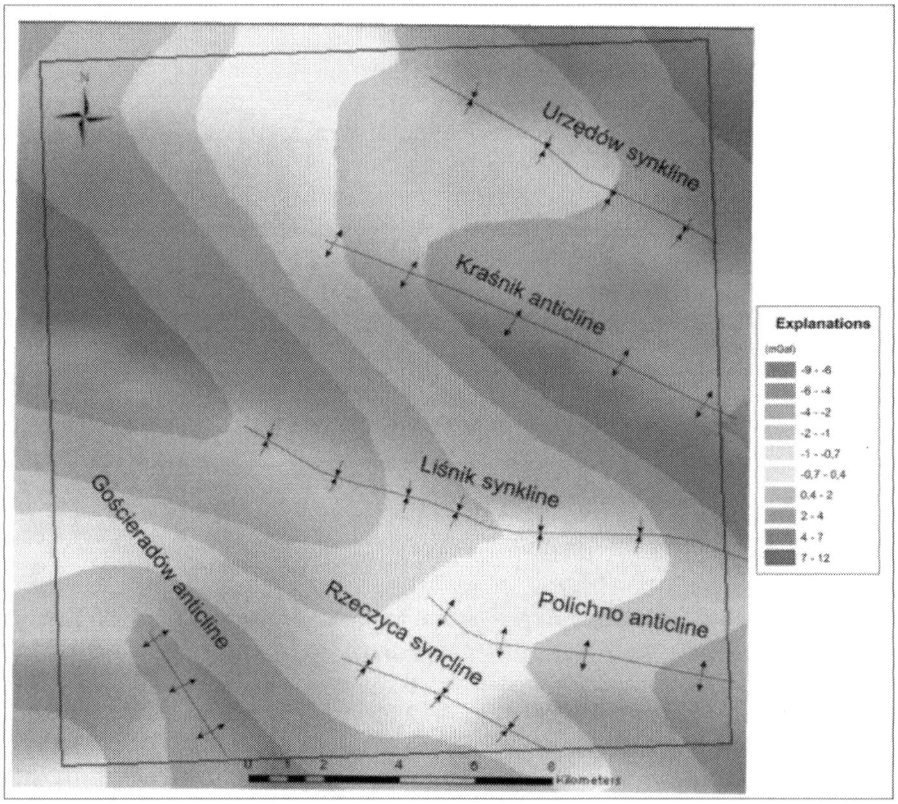

Fig. 2. The tectonic interpretation of the Bouguer anomaly.

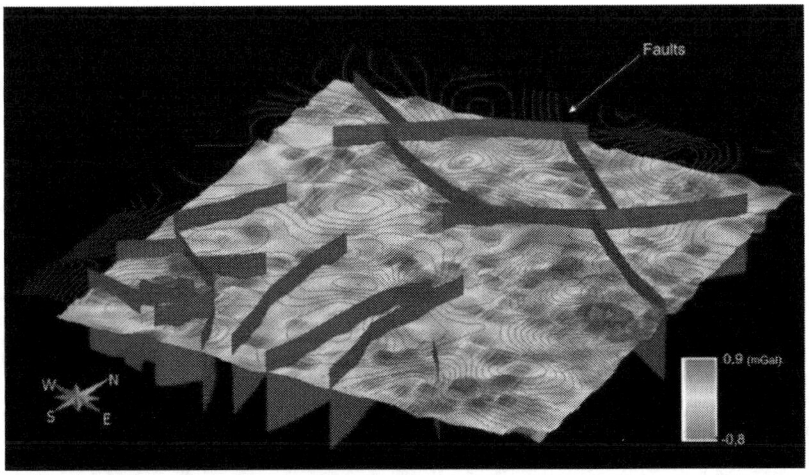

Fig. 3. The tectonic interpretation of residual anomalies.

4.2 Vertical Electric Sounding (VES)

The obtained RoT resistivity distribution image is a generalised image of the geological structure of the study area (Fig. 4). It is perpendicular to the axis of the Gościeradów anticline. This anticline was interpreted on the gravimetric anomalies image. Values from 2 to 2.8 ohm are visible in the resistivity values distribution for the core of the anticline – they may correspond to limestone rocks. Jurassic limestones are shown on the surface between sounding 24 and 29. The core of the anticline is surrounded by rocks with the resistivity values from 1.7 to 2 ohm on both sides. These are Cretaceous bedrocks. Stratigraphically younger marly bedrocks are deposited on the Cretaceous bedrocks from the NE side. Their resistivity values are within the range from 1.4 to 1.6 ohm. The shape of the anticline is asymmetric. It is cut with numerous faults which correlate with the faults interpreted from the residual gravimetric anomalies. Three regions with Quaternary formations were also determined on the Cretaceous formations. In the SW part of the section, there are high-resistivity sands, with values up to 2.6 ohm. In the NE part, two areas with low-resistivity loess formations were identified. Their resistivity values are within the range from 1.5 to 1.7 ohm.

4.3 Electrical Resistivity Tomography (ERT)

The ERT profile was made perpendicularly to the axis of the Gościeradów anticline structure. The profile shows high-resistivity Quaternary formations which are deposited unconformably on Jurassic and Cretaceous formations. Three boreholes – S.24, s.25 and s.26 were used to validate the resistivity values on the ERT profile. In the NE part of the profile, the interpretation shows sandy loess on loam sands. The resistivity values of the sandy loess reach 218 to 330 ohm. Loam sands have the resistivity of 500–600 ohm. Loam sands further in the profile transform into high-resistance sands with the values reaching almost 900 ohm. The maximum thickness of these sands is approx. 28 m (Fig. 5). The entire surface of the roof of the anticline is heavily eroded. This is evidenced by the significant differences in the elevations of the roof surface for over-Quaternary formations. Jurassic formations are at the lowest depth of approx. 0.5 m in the SW part of the profile. Their deepest deposition is found in the NE part of the profile. Jurassic and Cretaceous formations include: limestones, marly dolomites, sandstones and bedrocks. They are not separated on the ERT profile due to the lack of data required for their validation.

DOI: 10.1051/e3sconf/20172401005

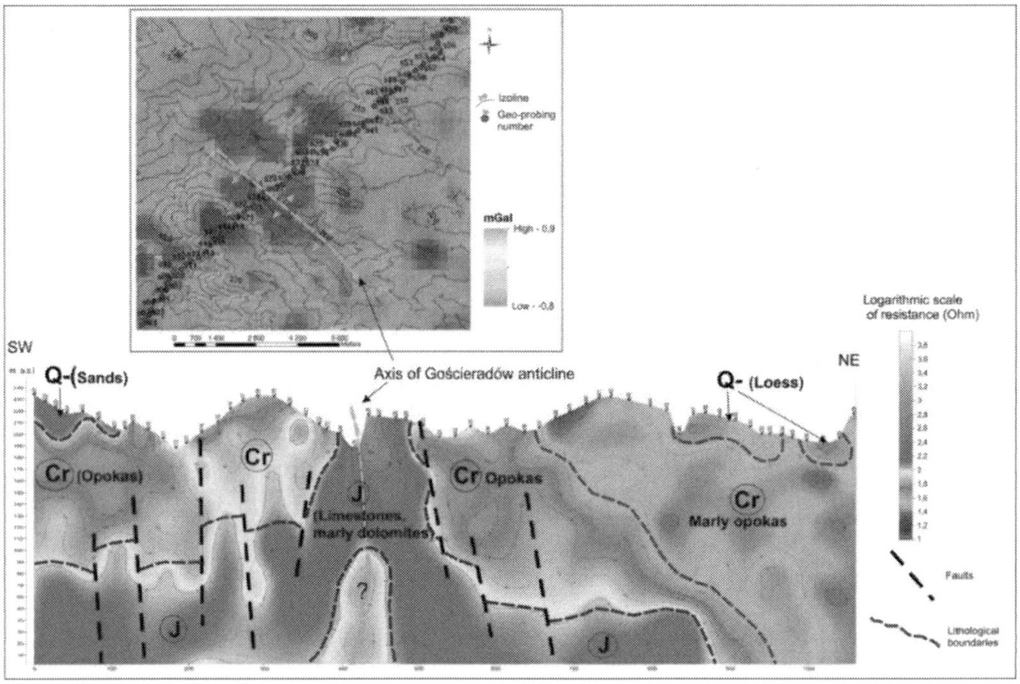

Fig. 4. Geological interpretation of the VES profile.

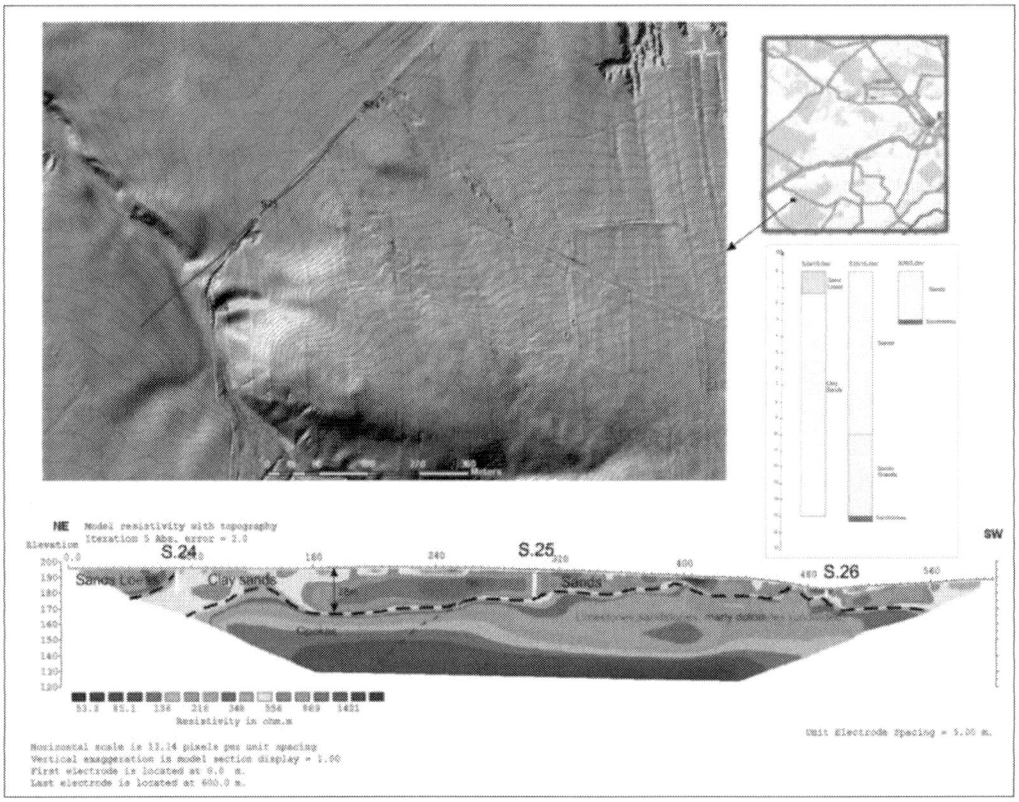

Fig. 5. Geological interpretation of the ERT profile.

4.4 ALS

The part of the area with the Gościeradów anticline was selected for the purposes of geomorphological photographic interpretation. The lithological lines of Jurassic limestone outcrops cut with two fault networks were interpreted on the ALS digital terrain model (Fig. 6). The direction of the first system is SW-NE and of the second – SE-NW. Side wings of the anticline are not readable in the topography due to being covered with Quaternary deposits.

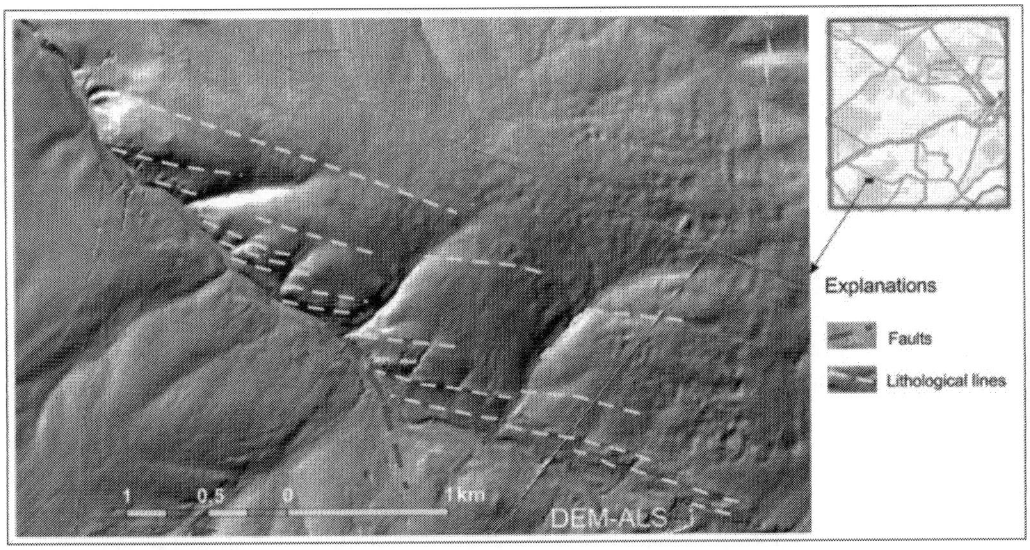

Fig. 6. Tectonic interpretation of the ALS model.

5 Discussion

The applied geophysical methods provided important information for the identification and interpretation of the geological structure of the study area. The ALS digital terrain model was also found to be useful – this type of information should be routinely used in geomorphological surveys.

As for the electrical resistivity methods, it should be noted that they have some limitations. They can be applied where there are relatively clear contrasts of electrical resistivity between adjacent geological formations. In the case of the VES method – its resolution decreases with depth. This is also true for the ERT – the resolution of the method decreases significantly at the depth of approx. 50 metres. It should be noted that the resolution of the Electrical Resistivity Tomography is however still much more precise than in the case of Vertical Electric Sounding. In the case of the Gościeradów anticline, it was evidenced by the fact the on the ERT profile, the high-resistivity sands were separated from the Jurassic rocks of the core of the anticline, while on the VES profile, the Jurassic rocks show similar resistivity values to the sands. Due to this fact, these lithological complexes could not be separated on the VES profile.

The analysis of the semi-detailed gravimetric photograph on the surveyed area applied to a number of structural information. It enabled the interpretation of the courses of fold structures and identify the directions of main faults. The analysis also showed significant variation in the density of rocks comprising the Kraśnik sheet area. It should be added that access to as many boreholes as possible is required for reliable interpretation of the gravimetric data. Unfortunately, the semi-detailed photograph does not seem to be sufficient for more precise tracking of rocks density in the scale of 1:50,000. It would be advisable to use the detailed gravimetric photograph for more precise determinations. The ALS numerical terrain model also proved to be useful for the geomorphological and geological interpretation of the topography of the areas largely covered by dense vegetation and forests. For this type of areas, aviation photography is not particularly useful.

Geophysical tests accompanied by the analysis of the ALS numerical model should be used comprehensively, in particularly selected ranges, as the information provided by different methods is complementary. The seismic, GPR, magnetic and conductometric methods could provide useful information for geophysical surveys applied to identify geological structures. Whereas, in the case of photogrammetric methods, the digital analysis of archival aviation photographs on photogrammetric stations should reinforce the results of studies, together with the application of ground-based laser scanning.

6 Conclusions

1. The ALS DTM method, electrical resistivity methods and the gravimetric method proved useful in the identification and interpretation of the geological structure of the Kraśnik region in the Lublin Upland.
2. The results of surveys enabled the identification of four anticlines and three synclines. The spatial distribution of the faults network was interpreted based on the archival geological data and the analysis of density differences in residual anomalies. Three directions were identified. In the southern part of the sheet, the dominant directions are N-S and W-E, while in the northern part, the most common directions are NE-SW and NW-SE.
3. The advantage of the gravimetric method is the possibility of spatial tracking of the distribution of geological structures, e.g. fold structures. This method enables the identification of fault zones which are often covered by overlaying younger formations (e.g. Quaternary).
4. The core of the anticline surrounded by rocks with lower resistivity values on both sides was interpreted from the distribution of resistivity values obtained by the VES method. The shape of the anticline is asymmetric. It is cut with numerous faults which correlate with the faults interpreted from the gravimetric data.
5. The ERT profile shows high-resistivity Quaternary formations which are deposited unconformably on Jurassic and Cretaceous formations. In the NE part of the profile, the interpretation shows sandy loess on loam sands. Loam sands transform into high-resistance sands with the values reaching almost 900 ohm. The maximum thickness of these sands is approx. 28 m. The entire surface of the roof of the anticline is heavily eroded. This is evidenced by the significant differences in the elevations of the roof surface for over-Quaternary formations.
6. In ERT imaging, the adopted measurement step – 10, 5 or 2 m – enabled tracking small geological structures which were not visible due to the large distances between consecutive VES soundings.
7. The lithological lines of Jurassic limestone outcrops cut with two fault networks were interpreted on the ALS digital terrain model. The direction of the first system is SW-NE and of the second – SE-NW. Side wings of the anticline are not readable in the topography due to being covered with Quaternary deposits.
8. Further work will be based on combining geological and geophysical data. As a result, three-dimensional geological models will be created.

References

1. T.L. Webster, J. Brendan Murphy, J.C. Gosse, I. Spooner, Can. J. Remote Sens, 2, 32, 173-193 (2006) doi: 10.5589/m06-017
2. M.H. Derron, M. Jaboyedof, Nat. Hazard Earth Sys. Sci. 9, 10, 1877–1879 (2010) doi: 10.5194/nhess-10-1877-2010
3. M. Kamiński, *Zastosowanie lotniczego skaningu laserowego i tomografii elektrooporowej w kompleksowych badaniach osuwisk: przykład z Pogórza Dynowskiego (Karpaty zewnętrzne),* Przegląd Geologiczny, 63, 410–417 (2015)
4. Z. Fajklewicz, *Grawimetria stosowana* (Wyd. AGH, Kraków 2007)
5. J. Kondracki, *Geografia regionalna Polski* (PWN, Warszawa 2002)
6. A.M. Żelichowski., *Obszar Radomsko-Lubelski,* [w:] *Budowa geologiczna Polski, IV, Tektonika, cz.* I (Państwowy Instytut Geologiczny, Warszawa 1974)
7. R. Griffin, *Atlas grawimetryczny Polski* (Państwowy Instytut Geologiczny, Warszawa 1949)
8. P. Krzywiec, W. Morawski., J. Twarogowski, *Zastosowanie metod meofizycznych do badań osadów kenozoicznych i zaburzeń glacitektonicznych na przykładzie południowej Warmii*, Prace Państwowego Instytutu Geologicznego CLXXXI, 13 – 36 (2004)
9. M.H Loke, *RES2DINV, rapid 2D resistivity and IP inversion using least-squares methods. User manual, advanced Geosciences* (Inc., Austin 1998)
10. M.H. Loke, *Tutorial: 2-D and 3-D electrical imaging surveys* (GeoTomo Software Ltd. 2012)
11. Z. Kurczyński, *Fotogrametria* (PWN, Warszawa 2014)
12. Cz. Królikowski, Z. Petecki, *Atlas grawimetryczny Polski* (PIG, Warszawa 1995)
13. O. Rosowiecka, Cz. Królikowski, K. Sobień, M. Gołębiewski, K. Lisowski, – *Opracowanie modelu rozkładu gęstości głównych jednostek geologicznych kraju* (PIG-NAG, Warszawa 2011)
14. K. Krassowska, *Kreda w okolicy Kraśnika - Zakrzewa rena podstawie głębokich otworów wiertniczych),* Przegląd Geologiczny 2, 25, 65-70 (1977)
15. K. Wyrwicka, *Stratygrafia, facje i tektonika mastrychtu zachodniej części Wyżyny Lubelskiej*, Kwartalnik Geologiczny 4, 24, 805-819 (1980)
16. J. Znosko, *Atlas Tektoniczny Polski* (Państwowy Instytut Geologiczny, Warszawa, 1998)

E3S Web of Conferences **24**, 01006 (2017) DOI: 10.1051/e3sconf/20172401006
AG 2017 - 3rd International Conference on Applied Geophysics

An application of the ground laser scanning to recognise terrain surface deformation over a shallowly located underground excavation

Elżbieta Pilecka[1,*], and *Dariusz* Szwarkowski[1]

[1]Politechnika Krakowska, Instytut Mechaniki Budowli, 31-155 Kraków, Poland

Abstract. In the Upper Silesian Coal Basin area, there are post-mining sites of shallow exploitation of metal ores and hard coal deposits that reveal discontinuous deformations. Most often, these areas are heavily urbanised and the appearing deformations may be dangerous to the existing building infrastructure. The work, described in this article, presents the results of the research, which aimed to rate the usefulness of laser scanning to recognize discontinuous deformations on surface areas located over shallow mining excavations. Two laser scanning measurements were taken over the course of a few months. The surface area images were compared to identify changes in its deformation, especially those areas located above mining excavations. The tests carried out by the laser scanning method showed that some of the identified discontinuous deformations could have been connected to the shallowly located mining excavations.

1 Introduction

Shallowly located mining excavations, which lose their stability over time, may have an influence on surface area deformations [1– 4]. Most often, different kinds of discontinuous deformations appear on the surface area: pit-holes, sinkholes, synclines, etc. [3,4] (Fig.1). In the Upper Silesian Coal Basin (USCB), these deformations pose a considerable risk to various building objects, including road infrastructure [5–7].

The effects of sinkhole processes that develop in mining terrains, are most often identified by different geophysical methods [8–13]. In the geological and mining conditions of USCB, the most frequently are used: gravimetric, seismic, electric, and georadar methods, among them borehole georadar [13].

The basic information on terrain surface deformations above mining areas can be obtained using a classic geodetic method [14]. More and more frequently, remote sensing methods are also used – satellite, or drone [15,16]. In some zones of surface deformation, ground laser scanning can also be used [17–21]. The advantages of the ground laser scanner are: uncomplicated measurement, time independence, and sufficiently accurate results [18]. Laser scanning has already proven to be useful in the reclamation of post-mining areas [19, 21, 22].

* Corresponding author: epilecka@pk.edu.pl

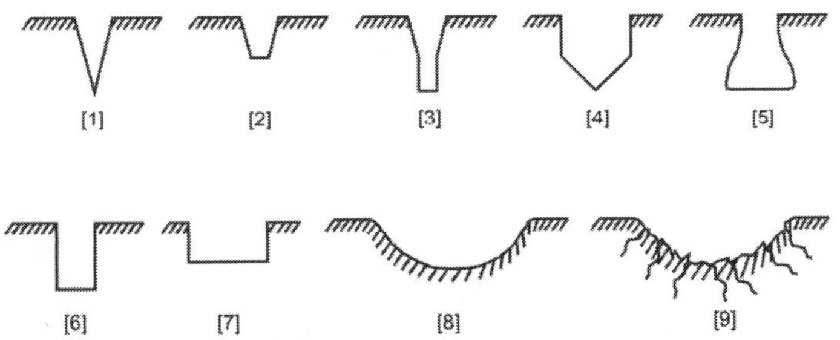

Fig.1. Kinds of discontinuous deformations of the terrain surface in view of geometries in the vertical section observed in the area of Upper Silesia Coal Basin [2 ,4].
Description: [1] – conical sinkhole, [2] – truncated sinkhole, [3] – conical-cylindrical sinkhole, [4] – cylindrical – conical sinkhole, [5] – bell sinkhole, [6] – cylindrical sinkhole, [7] – cuboid sinkhole, [8] – continuous syncline, [9] – discontinuous syncline.

The basic information on terrain surface deformations above mining areas can be obtained using a classic geodetic method [14]. More and more frequently, remote sensing methods are also used – satellite, or drone [15,16]. In some zones of surface deformation, ground laser scanning can also be used [17–21]. The advantages of the ground laser scanner are: uncomplicated measurement, time independence, and sufficiently accurate results [18]. Laser scanning has already proven to be useful in the reclamation of post-mining areas [19, 21, 22].

The work in this article presents the results of ground laser scanning in the post-mining, heavily urbanized area in USCB (Fig.2). The aim of the performed tests was to evaluate the usefulness of laser scanning to recognize discontinuous deformations on the surface, for areas located over shallow mining excavations at the depth of about 32 m. Two laser scanning measurements were taken over a few months. The surface area images were then compared to identify changes in deformation, especially those over mining excavations.

2 Research area characteristics

The research area is located in the southern part of Poland, in the USCB area (Fig.2), where intensive mining has taken place since the early Middle Ages. This area is mainly connected with shallow exploitation of zinc and lead ores and, to a lesser extent, iron and silver ores, and shallow and deep exploitation of hard coal deposits [7, 9, 10]. This research area is an example of a place that is presumably endangered by discontinuous deformations, due to the shallow exploitation of hard coal deposits.

Pleistocene fluvioglacial formations occur in depths up to 30.5 m in the studied area, in the form of varigrained sands, loamy sands, and sandy clays, as well as compact sandy clays with limestone crumbs. At deeper depths, there are carboniferous formations, in which mining excavations were made. Carboniferous formations in the top are represented by immature residual soil in the form of residual clay soil with shale interlayers, mudstones, sandstones, and hard coal inserts and deposits.

In the research area, at the depth of about 32 m, there are two main galleries in the direction close to a parallel and perpendicular to the beddings of the Carboniferous formations, excavated in the 19th and 20th century. The main galleries were crossed by roadways excavated in different coal seams. Additionally, the excavations were crossed with faults of throws of up to several meters (Fig.2). Due to such an arrangement of the excavations, a weak zones in rock mass were created, mainly in the area of excavation junctions, which presumably can create conditions for the development of sinkhole processes.

The surface of the studied area is a city park with big trees and shrubs. In the neighbourhood, there are also buildings, a pavement for pedestrians, and asphalt roads.

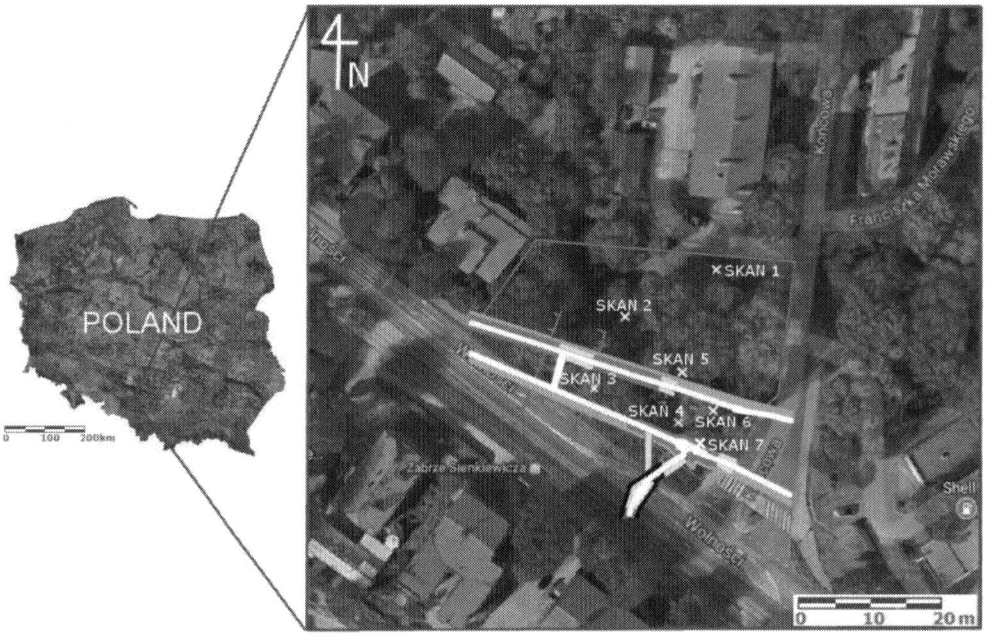

Fig. 2. Location of test stands, on the orthophotomap of the analysed area with marked excavations [10].

3 Methodology

The aim of the ground laser scanning is the most faithful mapping of the terrain surface. The measurements are based on sending and recording a laser beam, which is reflected from different objects [22] (Fig.3). The reflected signals, registered by the scanner, are converted into an electric signal and are then subjected to further digital processing.

The vertical lines on the signal envelope (Fig. 3) correspond to the sampling time. According to the waveform analysis and when it reached the receiver, calculations were made for the signals reflected in t1, t2,..., t4 times, for the subsequent reflecting objects. For the known light beam velocity and its run time from the emitter to the receiver, the distance between the scanner and the reflecting body is determined.

The wave emitted is characterised by high frequency – for Riegl VZ-4000 scanner f =122 kHz [22]. It allows for the scanning of several dozen to hundreds of thousands of measurement points in each measurement second. The obtained mapping of the scanned area is most often characterised by a dozen or more million points in the Cartesian space (x,y,z).

The data set, in the form of a point cloud, creates an image of the scanned space. An example of such an image is shown in Fig.3. It illustrates the research area described in this work. The measured space data was then archived for further processing. The factors influencing measurement quality deterioration include precipitations resulting in disturbances in the cloud of points. Precipitation causes the "noisiness" of the resulting image.

The difficulty in the laser recording processes is the vast amount of data. Therefore, the most essential fragments of the area are chosen for the analysis. The result of the ground laser scanner measurement is the final scan, formed as a result of combining data from various measurement stands (Fig.4). When processing the measurement data, trees and shrubs can create different difficulties. The registration of the changing position of leaves and branches during measurement increases the time of data processing. The removal of unwanted objects (particularly trees, shrubs, fences, etc.) is necessary and influences the simplicity of the scanned relief of the area.

The results presented in this work were obtained using a Riegl VZ-400 scanner. The maximum scan range for this scanner was 400 to 500 m [22], with an accuracy measurement of 5 mm. Measurements were taken in 'long scan mode', which has a maximum resolution. Scans from particular measurement stands were then linked to the reference scan in RiSCAN PRO.

The measurement by the ground laser scanner was carried out in two series on February 23, 2017 and June 1, 2017, covering the duration of spring melt. Measurements were made on seven stands which were positioned as shown in figure 2, marked by blue crosses. The measurement from February 23 made it possible to minimize the negative effects of the vegetation.

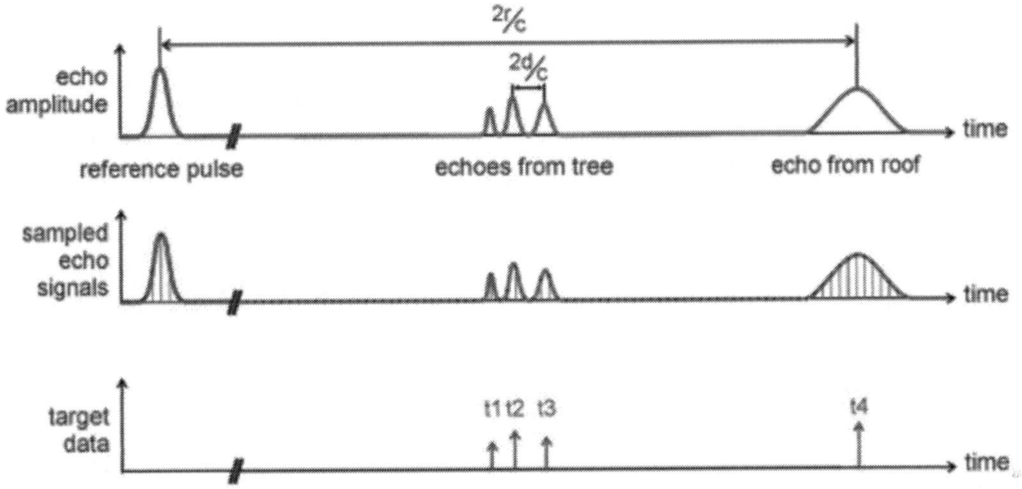

Fig. 3. Diagram of determining the position of objects in the measurement with a laser scanner. Description in text [23].

Fig. 4. An example of an image formed as a result of connecting scans from measurement stands in the research area.

4 Results and analysis

4.1 Measurement I

Laser scanning measurements were taken on February 23, 2017 which made it possible to develop a model of the surface area presented in Fig. 5. On the terrain surface model, mining excavations are denoted.

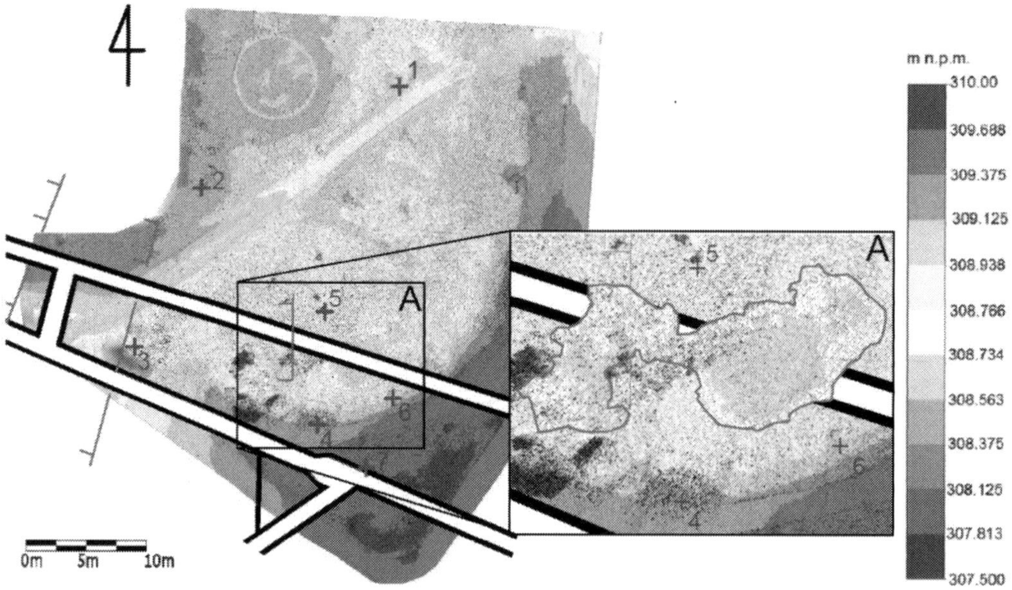

Fig. 5. The terrain surface model obtained from laser scanning on February 23, 2017.

The study area is characterised by a relatively flat surface of terrain. The measured elevations ranged from 307,500 to 309,688 m a.s. l. From the terrain surface image, it is evident that there are different zones of land subsidence of various sizes. The largest subsidence zone, limited by a blue line and marked as A in Fig. 5, is located over northern gallery. The zone is about 15 m long and 5 m wide. The local maximum land elevation in zone A differ by as much as 20 cm.

In zone A, from the south-east, there is also a distorted fragment between the area with the pavement for pedestrians and the neighbouring road, however the causes of such a deformation can be complex. Some of the subsidence zones in the entire studied area are beyond the possible impact of the excavations on the area surface. On the other hand, there is no anomalous subsidence visible in the north part of the studied area where the closest building is located.

4.2 Measurement II

Laser scanning measurements carried out on June 1, 2017 made it possible to develop a surface area model presented in Fig.6. The results of these measurements were very similar to the ones taken on February 23, 2017. Taking into consideration the results of both measurements, it should be stressed that local subsidence in the research area occurs in many places, not only on the mining sites.The largest of them and of the biggest local subsidence A', significantly exceeding the measurement error are marked over area between both galleries (Fig. 6).

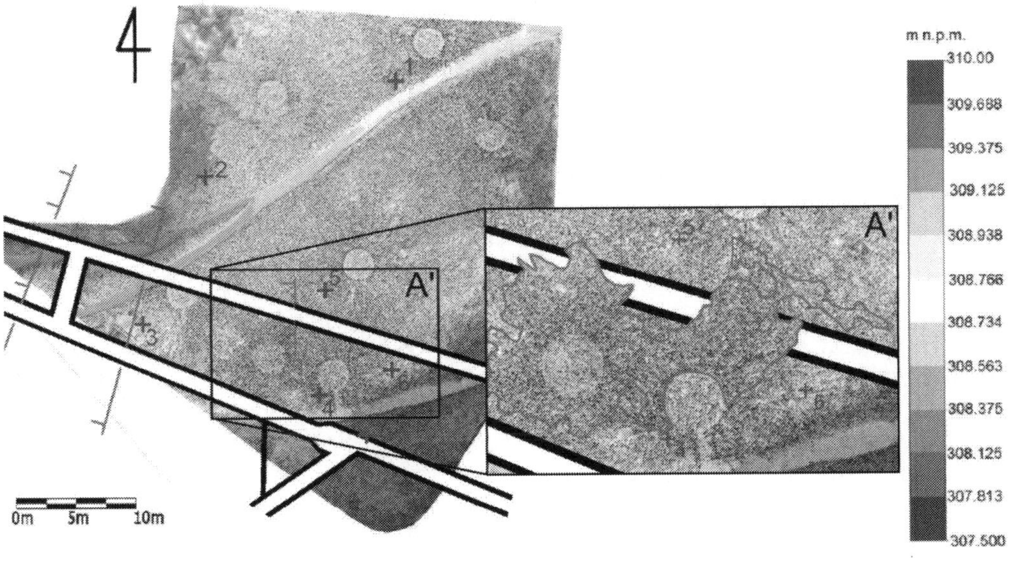

Fig. 6. Terrain surface model from laser scanning from June 1, 2017.

4.3 Differential model

The developed models of laser scanning in measurements I and II made it possible to develop a differential model of the studied area (Fig.7). The calculated differences in the ordinate values of the area surface are up to -2 cm. The differential zone, marked B1 in figure 7, occurs between the galleries, in the vicinity of the junction of south gallery with breakthrough. On the other hand, the B2 differential zone is north of the marked galleries, and from the available data, it is difficult to explain its origin.

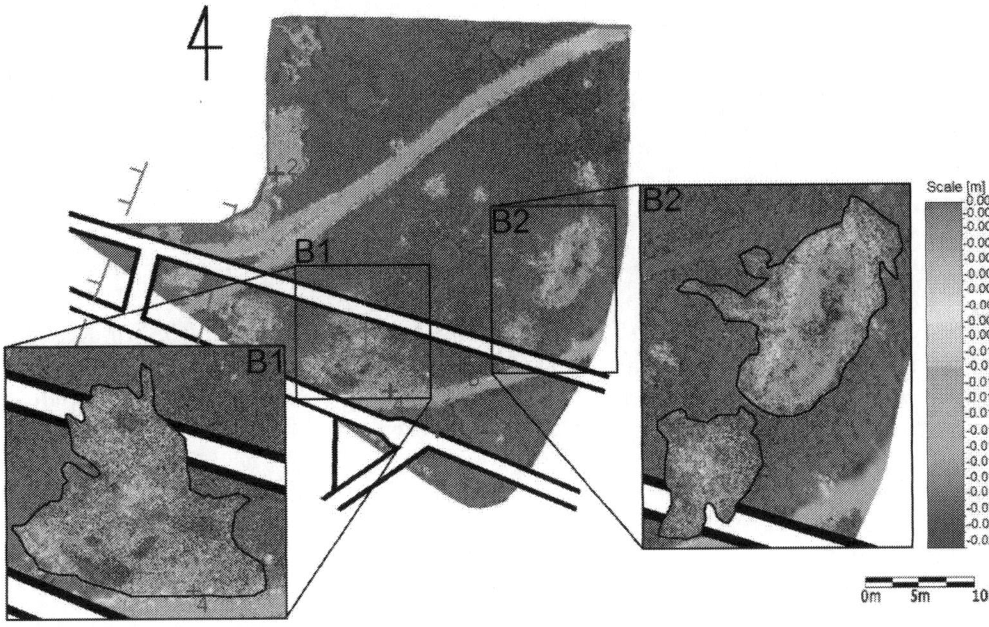

Fig. 7. Differential model of two laser scanning measurements on February 23, 2017 and June 1, 2017.

The results of differential calculations show that the biggest subsidence values are associated with A/A' zone in figures 5 and 6. This zone is located between the galleries, north of the junction of the south gallery and the breakthrough. The observed subsidence can have many causes, however what should not be excluded is the subsidence process developing in the ground over shallowly located, historical mining excavations. A small range of the subsidence zone indicates for a local phenomenon.

5 Summary

The work presents the results of laser scanning in the post-mining, heavily urbanised area of USCB, in order to evaluate the usefulness of the method for recognising discontinuous deformations on the terrain surface over shallowly located, historic mining excavations.

The results from two measurements taken over the course of several months proved the existence of local area subsidence zones, of values significantly exceeding that of the measurement error. Some of them occurred over the mining excavations. In zone B1, differential subsidence reached about 2 cm in a period of three months. Such a result may also be the effect of the defrosting of the ground surface, since the first measurements were made in February.

On the other hand, the differential values of the subsidence in B2 zone may have different causes, and on the basis of available data, it is difficult to explain this anomaly. However, the deformations found, are rather small and cannot be dangerous for the nearby buildings.

In order to examine the reasons of local subsidence in more detail, geophysical measurements properly chosen for the geological and mining conditions should be made. In the case of a geophysical anomaly, an inspection borehole would be required to verify this measurement.

The obtained results prove the usefulness of laser scanning to recognize discontinuous deformations on the terrain surface over shallowly located mining excavations.

This laser scanning method provides initial recognition for the probable occurrence of discontinuous deformations. These measurements are of great significance, especially in urban areas where other measurements are disrupted in different ways. They can also be used to assess the causes of damage to buildings, associated, for example, with mining activity.

References

1. B.N. Whittaker, D.J. Reddish, *Subsidence: occurrence, prediction and control* (Elsevier, Amsterdam 1989)
2. M. Chudek, *Geomechanika z podstawami ochrony środowiska górniczego i powierzchni terenu* (Wyd. Politechniki Śl., Gliwice 2002)
3. A. Goszcz, *Powstawanie zapadlisk i innych deformacji nieciągłych powierzchni na obszarach płytkiej eksploatacji górniczej*, Mat. Konf.: Szkoła Eksploatacji Podziemnej '96, 119-137 (CPPGSMiE PAN w Krakowie, Szczyrk 1996)
4. E. Popiołek, Z. Pilecki (eds.) *Ocena przydatności do zabudowy terenów zagrożonych deformacjami nieciągłymi za pomocą metod geofizycznych* (Wyd. IGSMiE PAN, Kraków 2005)
5. Z. Pilecki, *Methodology for A-1 motorway basement treatment effectiveness improvement by means of geophysical methods in the areas of metal ores shallow mining threatened with the sinkhole occurrence in the Upper Silesia*, Gospod. Surowcami Min. **3**, 25, 319-331 (2009)

6. A. Miłkowski, Z. Pilecki, K. Kłosek, M. Tondera, *Autostrada A1 zaprojektowana na „dziurawym" podłożu, Cz. 1*, Magazyn Autostrady: budownictwo drogowo-mostowe **3**, 104–112 (2010)
7. Z. Pilecki, A. Kotyrba, *Problematyka rozpoznania deformacji nieciągłych dla potrzeb projektowania konstrukcji drogowych na terenie płytkiej eksploatacji rud metali*, Prace Naukowe GIG III, 379-392 (2007)
8. H. Marcak, Z. Pilecki, *Some Geophysical and Geomechanical Remarks on Recognition Sinkhole Processes in Post-Mining Areas*, Proc. 12th European Meeting of Environmental and Engineering Geophysics, EAGE (Helsinki, Finland, 4 – 6 Sept. 2006)
9. Z. Pilecki, E. Popiołek, *Wpływ eksploatacji rud na zagrożenie powierzchni deformacjami nieciągłymi i jego badanie za pomocą metod geofizycznych*, Studia, Rozprawy, Monografie **84**, (Wyd. IGSMiE PAN, Kraków 2000)
10. Z. Pilecki, E. Pilecka, J. Pszonka, J. Stanisz, J. Morman, *Zagrożenia deformacjami nieciągłymi na obszarze górniczym Jaworzno III* (Wyd. IGSMiE PAN, Kraków 2012)
11. Z. Fajklewicz, Geophys. Prospect. 6, 34, 1233-1254 (1986) doi: 10.111/j.1365-2478.1986.tb00526.x
12. Z. Pilecki, *The role of geophysical methods in the estimation of sinkhole threat in the post-mining areas of shallow exploitation in the Upper Silesian Coal Basin, Poland*, Gospod. Surowcami Min. **3/1**, 24, 27-40 (2008)
13. T. Łątka, Pilecki Z. *Metoda oceny stopnia spękania ośrodka skalnego za pomocą georadaru otworowego*, Studia, Rozprawy, Monografie **200** (Wyd. IGSMiE PAN, Kraków 2017)
14. E. Popiołek, *Ochrona Terenów Górniczych* (Wydaw. AGH, Kraków 2009)
15. M. Kniotek, A. Leśniak, *Analiza związku sejsmiczności indukowanej z osiadaniami terenu na obszarze GZW z wykorzystaniem satelitarnych zobrazowań SAR*, Budownictwo Górnicze i Tunelowe **4**; 21, 36–41 (2015)
16. M. Graniczny, A. Piatkowska, E. Pilecka, *An applictaion of the remote sensing technology to the high energy seismic activity on the example of the Upper Silesian Coal Basin*, Polish Geological Institute Special Papers **20**, 92-98 (2005)
17. E. Pilecka, *Teledetekcja jako metoda analizy ruchów masowych*, Zeszyty Naukowe - Bulletin of the Mineral and Energy Economy Research Institute of the Polish Academy of Sciences **84**, 103-115 (2013)
18. Z. Muszyński, W. Milczarek, *Modelowanie przemieszczeń pionowych na podstawie danych z naziemnego skanera laserowego. Geoinformacja jako metoda ochrony przed geozagrożeniami*, 42-63 (Warszawa 2014)
19. E. Pilecka, D. Szwarkowski, *Wykorzystanie skanera laserowego RIEGL VZ-400 do badania deformacji na terenach górniczych GZW*, Budownictwo Górnicze i Tunelowe **4**, 21 (2015)
20. P. Gruchlik, *Naziemne skanowanie laserowe 3D, doświadczenia i perspektywy*, Przegląd Górniczy **5**, 71, 20-24 (2015)
21. M. Bazarnik, *Możliwości aplikacyjne techniki naziemnego skanowania laserowego 3D w rekultywacji terenów górniczych*, Zeszyty Naukowe - Bulletin of the Mineral and Energy Economy Research Institute of the Polish Academy of Sciences **94**, 149-160 (2016)
22. Riegl LMS, *Oprogramowanie Systemowe i Przetwarzania Danych RiSCAN PRO dla skanerów laserowych 3D firmy RIEGL LMS*, (Austria 2009)

Assessment of the impact of an overlying coal seam edge using seismic profiling of refracted P-wave velocity

Zbigniew Szreder[1], *Maciej* Barnaś[2*]

[1]JSW S.A. KWK Budryk, 43-178 Ornontowice, Poland
[2]Mineral and Energy Economy Research Institute, Polish Academy of Science, 31-261 Krakow, Poland

Abstract This paper presents the results of seismic profiling along the sidewalls of two headings of a longwall in a coal-seam at a depth of about 850 and 870 metres in a coal mine in the Upper Silesian Coal Basin, Poland. The seismic profiles were located in a zone of impact of the same overlying edge of the coal-seam located about 40 m above. This study was interesting from that point of view since there were no other geological and mining factors present which could disturb the impact of the coal seam edge. The profiling of refracted P-wave velocity changes was carried out according to the Dubinski method. This method is used for the assessment of relative stress in a coal seam in the side wall of the excavation. The results obtained on both seismic profiles are very similar, which demonstrates the small impact of the overlying edge of the coal seam in both headings of a longwall. It should be emphasised that at greater exploitation depths, the calculated reference velocity is less reliable than the measured reference velocity. Presumably, the method of calculating seismic anomaly requires updating under such conditions, but additional evidence should be collected.

1 Introduction

In underground coal mines, various types of hazards occur which affect the maintenance of the stability of excavations which are related to geological and mining factors [1-4]. Such hazards include various edges and remnants of already-exploited coal seams, which create excessive concentrations of stress in the rock mass and a simultaneously unfavourable impact on mining excavations [5-6]. Mining also causes the activation of faults, which creates concentrations of excessive stress [7-8]. The magnitude and extent of the impact of the unexploited edges and remnants of coal seams and activated faults can be measured using the seismic profiling method [3, 9]. This method is used for the assessment of relative stress in the coal seam in the longwall side of the excavation. The results achieved by seismic profiling are particularly useful in qualitative assessments of the stability of support-excavation systems and also rock bursts.

* Corresponding author: barnas@meeri.pl

There are many methods for evaluating the impact of various types of geological and mining factors on the behaviour of the excavation e.g. convergence, various borehole measurements, pressure sensors installed in the elements of excavation support, and geophysical measurements [10-12]. For this purpose, various types of analytical and numerical calculations are also used [13-17]. Under the conditions in Polish hard coal mines the method of seismic profiling [3, 9, 18-20] is commonly used.

Seismic profiling has been used to evaluate the effectiveness of destressing blasting. Tor et al. [21] presented an example of the control of the effectiveness of destressing blasting using seismic profiling in the Jas-Mos coal mine in the region of the 22a longwall in the 510/1 coal seam. The results of profiling have also been used to define a more detailed location for subsequent blastings. The possibility of seismic interpretation was also demonstrated, enabling one to determine the width of the fracture zone along the side-wall of the excavation.

Studies were also made of the profiling of changes in the attenuation coefficient and P-wave velocity in the coal seam along the side-wall of the headings [18, 22-23]. Szreder and Pilecki [18] concluded that the attenuation coefficient is more sensitive to changes of stress and deformation in the coal-seam compared to changes of refracted P-wave velocity. The value of an attenuation coefficient can vary many times more than the value of the seismic wave velocity. This issue has significant implications in practice because P-wave velocity increases are small at greater depths.

Seismic profiling in underground excavations was also carried out for different purposes in other geological conditions for example: in a salt body to estimate the width of the fracture zone in the side-walls of excavations [24] and also in tunnelling [25].

The paper presents the results of seismic profiling conducted in one of the hard coal mines in the Upper Silesian Coal Basin (USCB). The aim of these studies was to determine the anomalous changes in seismic P-wave velocity in the X0 coal seam in the sidewalls of A longwall headings, in the zone of impact of the overlying edge of the X1coal seam at depths of about 850 m and 870 m. The edge of the X1 coal seam was located about 40 m above the X0 coal seam. This study was interesting from that point of view since there were no other geological and mining factors present that could disturb the impact of the coal seam edge.

2 Theoretical assumptions of seismic profiling in coal seam

The basic parameter calculated from seismic profiling in a coal seam is the velocity of a refracted P-wave. This is the velocity of a wave propagating along an excavation in an elasticity zone (not disturbed by excavation) on the border with a plasticity zone (disturbed by excavation). The border between the zones of elasticity and plasticity may be illustrated by the use of a Ladanyi model [26], under conditions of interaction of excavation support with the surrounding rock mass (Fig.1).

The maximum of circumferential stresses occurs in a zone of elasticity on the border with a zone of plasticity (Fig.1). It should be assumption that the most favourable conditions for the propagation of a refracted wave (Fig. 2) occur in this part of the zone of elasticity and changes of P-wave velocity can correlate with changes of stresses. In practice, the border between zones of elasticity and plasticity has a transitional character and the propagation of seismic waves is more complex. A complicated character of wave field is visible on seismograms. A significant influence on the distortion of the wave pattern is the intensely variable width of the zone of plasticity along the excavation (Fig. 3). The determination of the onset of the refracted wave and its amplitude is frequently hindered under these conditions. This is demonstrated by the unreal results of inversion modelling which are often observed during an interpretation of the measurements of seismic profiling.

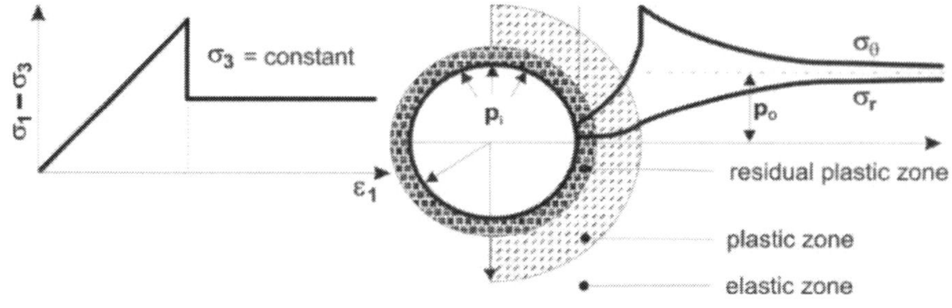

Fig. 1. Model of behaviour of a rock mass around an excavation on the basis of Ladanyi [26] σ_1 – major principal stress, σ_3 - minor principal stress, σ_θ – tangential stress, σ_r – radial stress, ε_1 – major principal strain, p_o – virgin stress, p_i – lining load [18].

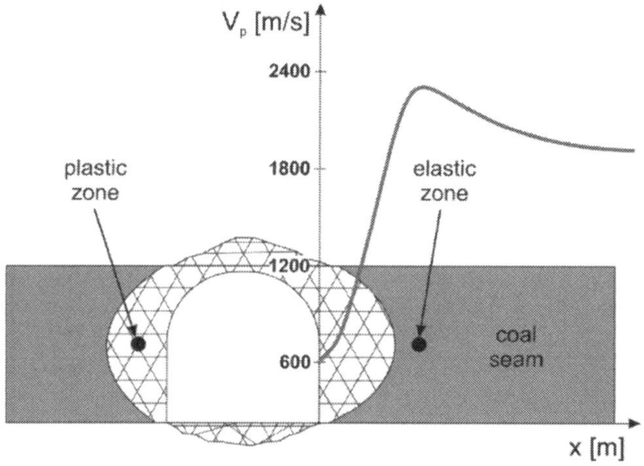

Fig. 2. An example of changes in P-wave velocity in the side-wall of the excavation [18].

The determination of the onset of a P-wave in a heterogenous medium is difficult due to the complicated wave field in a plasticity zone in close proximity to the sidewall of excavations. Many types of waves may form in such conditions: direct, reflected, diffracted, refracted also along the roof and floor of the coal seam or channel waves.

3 Methodology

Profiling of the refracted P-wave velocity is performed according to the Dubiński method [9] updated in work of Dubiński and Konopko [3]. In practice, interpretation of the profiling is carried out with the help of refraction algorithms for a two-layer model.

Depending on the conditions of refracted P-wave propagation in a coal-seam, the length of spread for 24 geophones may reach up to 115 metres adjusted to the energy of the wave source. The geophone interval should be assumed to be from 2 to 5 metres depending on the possibility of identifying refracted P-waves. The seismic wave is excited with the use of a stroke with a sledgehammer. For improvement of the signal to noise ratio, at least six fold stacking should be applied. The seismic records are legible even in longer sections in the case of a low level of seismic noise and high amplification in the test equipment of the order of over 100 dB. The geophone may be installed in a different way, in addition to which the contact with the rock mass is to secure distinct and non-deformed onsets of refracted waves. The geophones are usually fixed on short anchors a few centimetres long in not intensively fractured and not separated fragments of hard coal. This ensures that the wave image is useful enough for an interpretation. The recording time and signal sampling have always to be tested

DOI: 10.1051/e3sconf/20172401007

in a specific investigation site. On the basis of previous experiments, one should select a sample interval of 0.125 ms and record length of 0.5 sec.

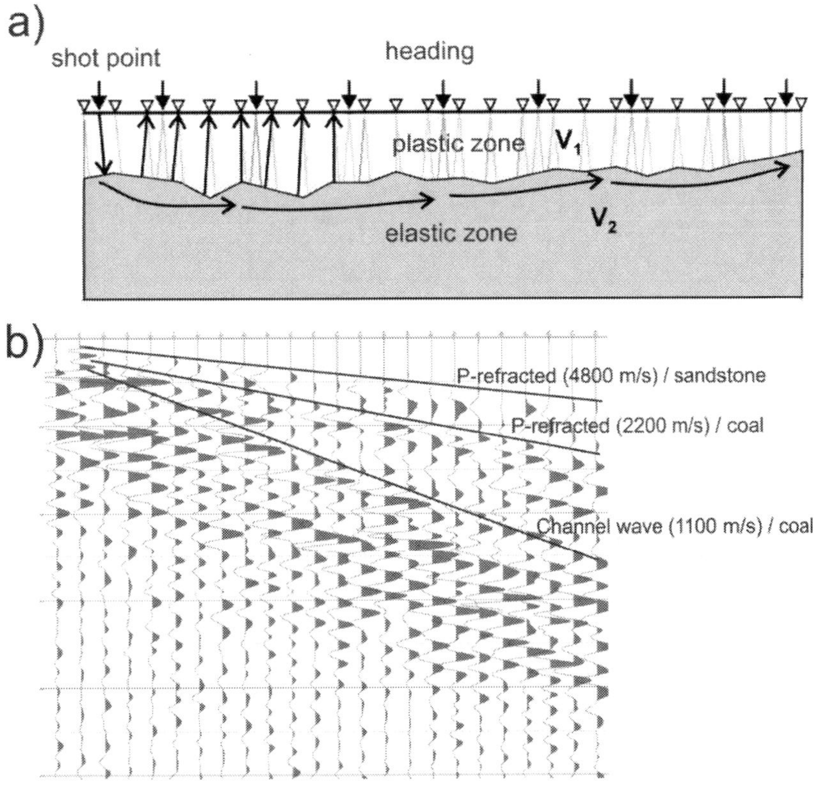

Fig. 3. Seismic profiling in a coal seam along an excavation side-wall: (a) refracted P-wave wave propagation (V2 velocity); (b) example of wave pattern [18].

In the interpretation stage, a two-layer model of a medium consisting of a fracture zone (plastic zone) and a solid zone (elastic zone) is assumed. The reciprocal travel times method is used to calculate the velocity model and width of the fracture zone. The velocity model is corrected by the reverse analysis method. By changing the position of the seismic boundaries of the model, the calculated hodographs are adjusted to the observations. The accuracy of the calculations is verified by minimising the average square error. Figure 4 shows the successive stages of seismic profiling.

In the studies, two modules of Geometrics SeisImager software were applied, Pickwin and Plotrefa, which are shown on the diagram in Figure 4. The Pickwin is used to identify the first breaks of P-wave. Due to an unfavourable signal-to-noise ratio and the complexity of the recorded waveform (interference of longitudinal and transverse waves), frequency filtering was applied in the Pickwin software. In the next stage, the results from Pickwin were used in the Plotrefa software for data interpretation, according to the procedures outlined in Figure 4.

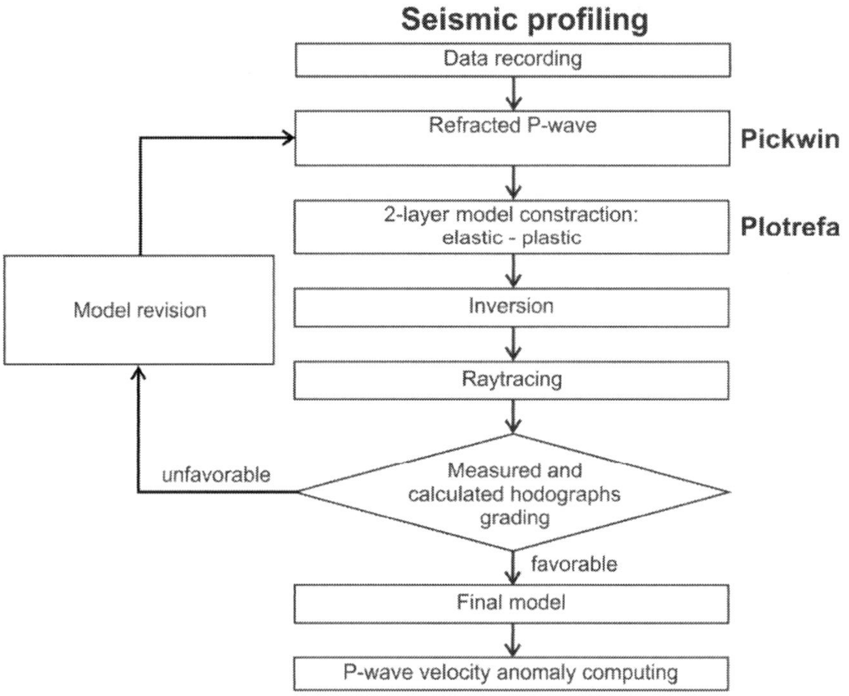

Fig. 4. Block diagram of seismic profiling in the mining excavation.

From this, the refracted P-wave velocity in the coal seam was calculated along the profile. In the final stage, velocity changes were compared with the reference velocity determined from the Dubinski empirical equation [3]:

$$V_p^0 = 1200 + 4.83H^{0.76} \tag{1}$$

where:

V_p^0 – reference velocity [m/s]

H – depth of survey area [m].

The reference velocity V^0 and the measured velocity V^R were used to calculate the percentage anomaly A^0 according to the formula:

$$A^0 = \frac{V^R - V_p^0}{V^R} \cdot 100\% \tag{2}$$

Sometimes, the anomaly used to be calculated on the basis of measured velocity in the undisturbed zone instead of using the calculated reference velocity. The calculated anomaly of the velocity is applied to evaluate the relative stress state based on the seismic scale, determined for USCB conditions (Table 1 and 2).

Table 1. Seismic scale of relative stress increase under USCB conditions [3].

Degree of relative stress increase	Scale of relative stress increase	Seismic anomaly [%]	Probable increase in relative stress [%]
0	lack/very low	below 5	below 20
1	low	5-15	20-60
2	medium	15-25	60-140
3	high	above 25	above 140

Table 2. Seismic scale of relative stress decrease under USCB conditions [3].

Degree of stress decrease	Scale of relative stress decrease	Seismic anomaly [%]	Probable decrease in relative stress [%]
0	lack/very low	above -7.5	below 25
1	low	-7.5 ÷ -15	25 ÷ 55
2	medium	-15 ÷ -25	55 ÷ 80
3	high	below -25	above 80

4 Location and geological and mining conditions

Seismic measurements were conducted in the region of a planned A longwall in the X0 coal seam (Figure 5a). The seismic profile I - I ' was carried out in the A1 heading at a depth of about 870 m. The seismic profile II - II' was carried out in the A2 heading at a depth of about 850 m. Both seismic profiles were located in the impact zone of the overlying edge of the X2 coal seam, lying about 40 m above the X0 coal seam. Both the seismic profiles were 135m long.

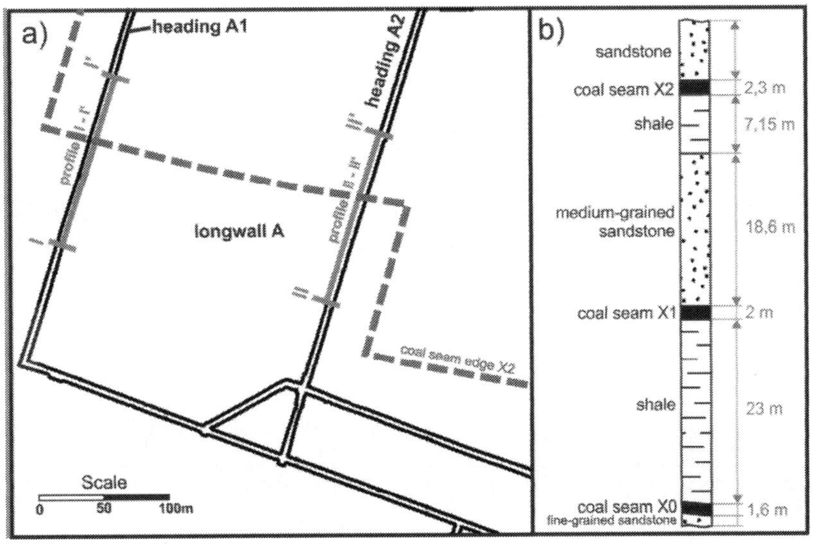

Fig. 5. (a) Location of the seismic profiles in the region of the A longwall in the X0 coal seam, with the edge of the X2 coal seam indicated; (b) geological profile from the region of the A longwall.

In the research area, the X0 coal seam has a mean thickness of about 1.60m (Figure 5b). The slope of the seam is approximately 6° NW. A shale layer of 23 m in thickness occurs as the immediate roof of the X0 seam. A medium-grained sandstone layer of 18.6 m in thickness and a layer of shale of 7.15 m in thickness occur above the X1 coal seam. Coal seam X2 is located approximately 40 m above seam X0. A fine sandstone layer of 18.6 m in thickness occurs below the X0 coal seam.

5 Results and analysis

A diagram of the changes of refracted P-wave velocity on both profiles I – I' and II – II' in a region of the A longwall in the X0 coal-seam is presented on Fig. 6. The error of the measured velocity did not exceed 50m/s and the velocity differences on both profiles were greater than the measurement error.

 DOI: 10.1051/e3sconf/20172401007

5.1 Profile I- I'

Throughout the entire length of the profile I – I', the P-wave velocity varies from 2180 m/s to 2320 m/s. On the basis of velocity changes, two sections can be distinguished: first from 0 to about 90 m, second from about 90 to 135 m of the profile length. In the first section, the velocity of the P-wave changes from 2220 m/s to 2320 m/s. The maximum velocity occurs at about 80 m. The calculated anomaly equal to A1 = 7.8% for a reference velocity of 2140 m/s. The level of the anomaly indicates a low increase of relative stress between 20% and 60%.The influence range of this anomaly is estimated to be from about 68 m to 90 m. Assuming that on the initial section of the profile, measured wave velocities P are not disturbed by an edge influence and other geological and mining factors, it can be assumed that the measured reference velocity is 2220 m/s. With respect to this velocity, the anomaly A1' is equal to 4.3%, which corresponds to a very low increase in relative stress up to 20%.

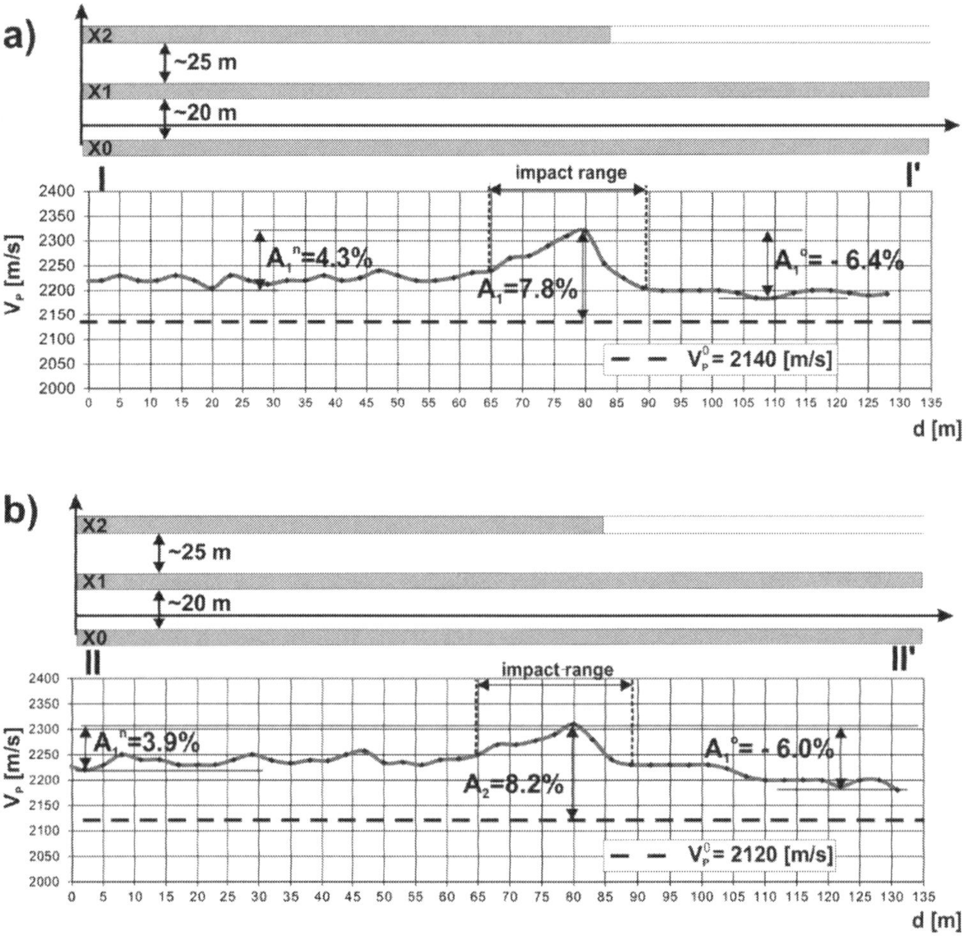

Fig. 6. Changes of refracted P-wave velocity in the X0 coal seam in the influence zone of the overlying edge in the X2 coal seam: a) seismic profile I – I' in drift A1, b) seismic profile II – II' in drift A2.

The value of anomaly A1', according to the assumption of the Dubinski classification, seems to be more probable than the anomaly A1, because it was calculated from the measurement data. The results indicate a low impact of the edge of the X2 coal seam on the X0 coal seam in the heading A1.

On the second section of the profile I - I' under the gobs of the X2 coal seam, the P-wave velocity varies from 2180 to 2230 m/s. The smaller P-wave velocity compared to the first section indicates the destressing impact of the gobs of the X2 coal seam. However, the destressing impact of these gobs is small, because the calculated anomaly compared to the maximum velocity value of 2320 m/s is equal to A1" = -6.4%. The small destressing impact of the gobs of the X2 coal seam corresponds to the small increase of relative stresses induced by the edge of X2 coal seam.

5.2 Profile II- II'

On the profile II – II', the P-wave velocity varies from 2180 m/s to 2320 m/s, which is similar to profile I – I'. On the basis of the velocity changes, two sections can be distinguished: first from 0 to about 90 m, second from about 90 to 135 m of the profile length. In the first section, the velocity of the P-wave changes from 2220 m/s to 2310 m/s. The maximum velocity occurs at about 80 m. The calculated anomaly is equal to A2 = 8.2% for a reference velocity of 2120 m/s. The level of the anomaly indicates a low increase of relative stress of between 20% and 60%.The influence range of this anomaly is estimated to be from about 72m to 97m. Assuming that on the initial section of the profile, measured wave velocities P are not disturbed by an edge influence and other geological and mining factors, it can be assumed that the measured reference velocity is 2220 m/s. With respect to this velocity, the anomaly A2' is equal to 3.9%, which corresponds to a very low increase in relative stress up to 20%.

On the second section of the profile II - II' under the gobs of the X2 coal seam, P-wave velocity varies from 2180 to 2230 m/s, identical to that on profile I – I'. The destressing impact of the gobs of the X2 coal seam is also small, since the calculated anomaly, compared to the maximum velocity value of 2320 m/s, is equal to A2" = -6.0%.

The results obtained on both seismic profiles are very similar, which demonstrates the small impact of the edge of X2 coal seam on the X0 coal seam in both headings of the A longwall. The compatibility of the results on both profiles indicates the great repeatability of the seismic profiling surveys. It should be underlined that, at greater exploitation depths, the calculated reference velocity is less reliable than the measured reference velocity.

6 Conclusions

Under the conditions of multi-seam exploitation in underground coal mines, excavations are often located in the zone of excessive impact of various types of edges and remnants of unexploited coal seams. Seismic profiling is one of the fundamental methods, which allow the assessment of such impacts directly in the hazard zone. On the basis of the results of seismic profiling presented in the paper, the following conclusions can be formulated:

1. Seismic profiling provides reliable information about P-wave velocity changes in the coal seam along the side-wall of the excavation. The results of surveys conducted in two different impact zones of the same edge of a coal seam presented similar values of refracted P-wave velocity and calculated seismic anomaly.
2. At greater depths, the calculated reference velocity may vary from the measured velocity. For the calculation of the anomaly, measured velocity can give more valuable information.
3. In Polish hard coal mines exploitation takes place at increasingly greater depths, even below 1300 m. The method of calculating a seismic anomaly can be presumed to require updating under such conditions. To achieve this goal, more experimental results should be obtained.

References

1. Z. Burtan, *The influence of natural hazards on occupational safety in Polish collieries,* Proc. of Int. Mining Forum on mine safety and efficient exploitation facing challenges of the 21st century, 81-90 (2010)
2. Z. Burtan, Z. Rak, J. Stasica, *Priorities in the development of mining systems in the mining sector in Poland,* Gospod. Surowcami Min. **24**, 1, 185-200 (2008)
3. J. Dubiński, W. Konopko, *Tąpania – ocena, prognoza, zwalczanie* (Wydawnictwo GIG, Katowice, 2000)
4. J. Dubiński, Z. Pilecki, W. Zuberek, *Badania geofizyczne w kopalniach – przeszłość, teraźniejszość, i zamierzenia przyszłość* (IGSMiE PAN, Kraków 2001)
5. M. Kwaśniewski, J. Wang, *Numerical analysis of mining-induced deformations of the strata*, Proc. of The School of Underground Mining '96, 153-187 (in Polish) (Centrum PPGSMiE PAN, Kraków 1996)
6. Z. Pilecki, *Modelowanie numeryczne pola naprężenia w górotworze naruszonym wielopokładową eksploatacją węgla kamiennego w warunkach silnego zagrożenia sejsmicznego*, Zeszyty Naukowe - Bulletin of the Mineral and Energy Economy Research Institute of the Polish Academy of Sciences **80**, 93-102 (2011)
7. J. Dubiński, K. Stec, *Relationship between focal mechanism parameters of mine tremors and local strata tectonics*, Proc. of the 5th International Symposium on Rockbursts and Seismicity in Mines (January 2001)
8. Z. Burtan, A. Zorychta, J. Cieślik, D. Chlebowski D., Arch. Min. Sci. **59**, 3, 691 – 704 (2014) doi: 10.2478/amsc-2014-0048
9. J. Dubiński, *Sejsmiczna metoda wyprzedzającej oceny zagrożenia wstrząsami górniczymi w kopalniach węgla kamiennego* (Prace GIG, Seria dodatkowa, Katowice 1989)
10. B.H.G. Brady, E.T. Brown, *Rock mechanics for underground mining,* (University Press, Cambridge 1993)

A. Kidybiński, *Podstawy geotechniki kopalnianej* (Katowice 1982)

11. R. Czarny, H. Marcak, N. Nakata, Z. Pilecki, Z. Isakow, Pure Appl. Geophys. **173**, 6, 1907-1916 (2016) doi: 10.1007/s00024-015-1234-3.
12. E. Hoek, P.K. Kaiser, W.F. Bawden, *Support of underground excavations in hard rock* (1995)

A. Tajduś, M. Cała, K. Tajduś, Studia Geotechnica et Mechanica **38**, 1, 109 - 121 (2016) doi: 10.1515/sgem-2016-0012

13. Z. Pilecki, *Metoda oceny zachowania się masywu skalnego wokół wyrobiska podziemnego*, Studia, Rozprawy, Monografie **59** (IGSMiE PAN, Kraków 1999)
14. Z. Pilecki, *Dynamic analysis of mining tremor impact on excavation,* Proc. Int. FLAC Symp. on Numerical Modeling in Geomechanics, Detournay & Hart, 397-400 (Minneapollis, USA 1-3 Sept. 1999)
15. Z. Pilecki, G. Mutke, *Oddziaływanie wstrząsu górniczego w strefie bliskiego pola falowego na wyrobisko chodnikowe - analiza numeryczna,* Mat. X Międzyn. Konf. Tąpania, Główny Instytut Górnictwa, 223-232 (Ustroń, 12-14 Nov. 2003)
16. Z. Szreder, Z. Pilecki, J. Kłosiński, *Effectiveness of recognition of exploitation edge influence with the help of profiling of attenuation and velocity of seismic wave,* Gospod. Surowcami Min. **24**, 2, 215-226 (2008)
17. Ł. Wojtecki, *Wykorzystanie metod sejsmicznych w wersji kinematycznej do określania stanu naprężeń w górotworze podczas eksploatacji w Kopalni Węgla Kamiennego „Bielszowice"*, 578-597 (Prace Naukowe GIG, Górnicze Zagrożenia Naturalne, Katowice, 2011)

18. Ł. Wojtecki, G. Dzik, *Profilowanie sejsmiczne ociosów chodników węglowych w wersji kinematycznej i tłumieniowej*. Mechanizacja i Automatyzacja Górnictwa **50**, 4, 29 – 37 (2012)

A. Tor, P. Chmiel, Z. Pilecki, A. Jakubów, R. Skatuła, Z. Szreder, J. Kłosiński, *Kontrola efektywności strzelań torpedujących za pomocą profilowań sejsmicznych*, Bezpieczeństwo Pracy i Ochrona Środowiska w Górnictwie, Miesięcznik WUG **142**, 6, 62-64 (2006)

19. Z. Pilecki, Z. Szreder, *Możliwości wykorzystania współczynnika tłumienia do oceny stanu naprężenia i deformacji za pomocą profilowań sejsmicznych w warunkach dużych głębokości,* Prace Naukowe GIG **4**, 197-203 (2007)

20. Z. Szreder, Z. Pilecki, J. Kłosiński, *Comparison of Profiling Results of Attenuation and Velocity of Refracted P-wave in Coal-seam,* Proc. 14th European Meeting of Environmental and Engineering Geophysics, EAGE (Krakow 15-17 September 2008)

21. J. Ślizowski, Z. Pilecki, K. Urbańczyk, E. Pilecka, L. Lankof, R. Czarny, Adv. High Energy Phys., 1-12 (2013) doi: 10.1155/2013/461764

22. Z. Majcherczyk, Z. Pilecki, E. Niedbalski, E. Pilecka, M. Blajer, J. Pszonka, *Wpływ warunków geologiczno-inżynierskich i geotechnicznych na dobór parametrów obudowy wstępnej tunelu drogowego w Lalikach*, Gospod. Surowcami Min. **28**, 1, 103-124 (2012)

23. B. Ladanyi, *Use of the long-term strength concept in the determination of ground pressure on tunnel linings*, Proc. of the Third Congress of the Int. Soc. for Rock Mech., II, B, 1150-1156 (Denver, 1974)

Wireless acquisition of multi-channel seismic data using the Seismobile system

Zbigniew Isakow[1,*]

[1]Institute of Innovative Technologies EMAG, 40-189 Katowice, Poland

Abstract. This paper describes the wireless acquisition of multi-channel seismic data using a specialized mobile system, Seismobile, designed for subsoil diagnostics for transportation routes. The paper presents examples of multi-channel seismic records obtained during system tests in a configuration with 96 channels (4 landstreamers of 24-channel) and various seismic sources. Seismic waves were generated at the same point using different sources: a 5-kg hammer, a Gisco's source with a 90-kg pile-driver, and two other the pile-drivers of 45 and 70 kg. Particular attention is paid to the synchronization of source timing, the measurement of geometry by autonomous GPS systems, and the repeatability of triggering measurements constrained by an accelerometer identifying the seismic waveform. The tests were designed to the registration, reliability, and range of the wireless transmission of survey signals. The effectiveness of the automatic numbering of measuring modules was tested as the system components were arranged and fixed to the streamers. After measurements were completed, the accuracy and speed of data downloading from the internal memory (SDHC 32GB WiFi) was determined. Additionally, the functionality of automatic battery recharging, the maximum survey duration, and the reliability of battery discharge signalling were assessed.

1 Introduction

Geophysical methods have been demonstrated to be effective tools in imaging the shallow subsurface beneath transport routes and assessing mechanisms of subsurface deformation [1–8]. Many studies have investigated these methods. Among the various geophysical methods, the most applicable for designing new transport routes and rehabilitating existing ones are GPR, seismics, and electrical resistivity [8]. In this context, the geological and engineering conditions of the shallow subsurface are analysed to identify structurally weak zones. Weak zones are identified on the basis of variations in the physical properties of the medium, zones of irregular groundwater flow (including leaching zones), changes in the ground water level caused by drainage or irrigation infrastructure, organic soil and quick-sands, landslide zones, tectonics, the occurrence of voids and fractures due to mining and karst areas [8–10].

Interest in mobile landstreamer seismic systems has increased in recent years [8,11]. The Seismobile system was developed by combining seismic landstreamers and GPR profiling [8,12]. The Seismobile system enables wireless geophysical subsurface profiling, including

* Corresponding author: z.isakow@ibemag.pl

refraction, reflection, MASW (Multi-channel Analysis of Surface Waves), and GPR. The main system component is the seismic landstreamer, which permits faster seismic measurements under conditions difficult for geophone mounting, such as concrete or asphalt road surfaces. The Seismobile landstreamer is a mobile construct that enables 1 – 4 woven belts to be towed with a total of 96 geophones with the accompanying modules for survey geometry measurement. More information regarding the Seismobile system can be found in [8, 11-15].

The known methods and systems for survey data collection, discussed in [11], are based on an array of streamers with electrodynamic geophones, accelerometers, or mems connected by wired analogue transmission to multi-channel (usually 24 channel) loggers transported on a moving vehicle used as a platform for translocation of the streamers. The use of analogue transmission with large measuring sets (up to 100 sensors) hampers efforts to obtain dynamic recording, due to the potential for electromagnetic interference, which is important when recording weak seismic signals. The cable connections of traditional seismic systems are exposed to damage, and the equipment require time-consuming cables coupling. The Seismobile system eliminates these inconveniences and reduces the time required for system setup by providing a wireless data acquisition platform that eliminates the need for the wired analogue transmission of sensor signals. Brodic et al. [16] have developed an innovative landstreamer system based on 3C MEMS (Micro-Electro-Mechanical System) accelerometers combined with wireless units that are GPS time stamped with nanosecond accuracy. Moreover, the broadband (theoretically 0–800 Hz) MEMS-based sensors enable high-resolution imaging.

The Seismobile system digitally measures data that are recorded on the system's internal Flash memory, while radio transmission is used to configure and control the status of system components, impacting the accuracy of the recording time base synchronization and the initial validation of seismic recording.

2 Structure, specifications, and operation of the system and its components

The schematic of the 3D version of the Seismobile system is shown in Figure 1. The construction of the Seismobile system permits towing of up to four measuring lines with geophones and the GPR antenna. The seismic and GPR control units are placed on a vehicle and work independently. The positions of the seismic detectors and the GPR antennas are synchronized by GPS.

The main part of the system is comprised of four measuring landstreamer belts, with each measuring approximately 100 m long, that enable a maximum of 96 uni- or tri-axial geophone sensors (5). Each sensor works with a measuring module (6) that samples the signal from the sensor and performs the analogue to digital conversion with a dynamic range of 130 dB and a resolution of 32 bits. Sampling is executed precisely by synchronizing the clocks in the modules (6) with the GPS receivers. The clock within the triggering measurements module (10), assisted by the accelerometer sensor (11) that identifies a moment of the seismic wave excitation, is synchronized with its GPS receiver (12). The synchronous operation of the measuring modules (6) and the control module (10) permits precise time-based synchronization during the digital recording of seismic waves produced by a pile driver source or a hammer. The waveforms (5) recorded in the internal memory of the modules (6) after each wave excitation (7) are wirelessly transmitted to the central processing unit (14), where they are visualized and verified. After each wave excitation, the coordinates of the pile driver and the acceleration of the induced waveform are recorded in the module (10), while the waveforms of the landstreamer sensors (6) are recorded in the memory of a separate set of modules (6). The landstreamer modules (6) are automatically numbered as they are

DOI: 10.1051/e3sconf/20172401008

fastened to the bases on the landstreamer belts, which are associated with the coordinates from the GPS receiver installed on the ST vehicle and provide unambiguous spatial geometry for the sensor deployment. After survey acquisition, the MP measurement modules are placed in the docking station SD, where their batteries are charged, and the data are downloaded via WiFi. More details on the construction of system components are provided in [13, 14]. An example of the seismic acquisition parameters is presented in Table 1.

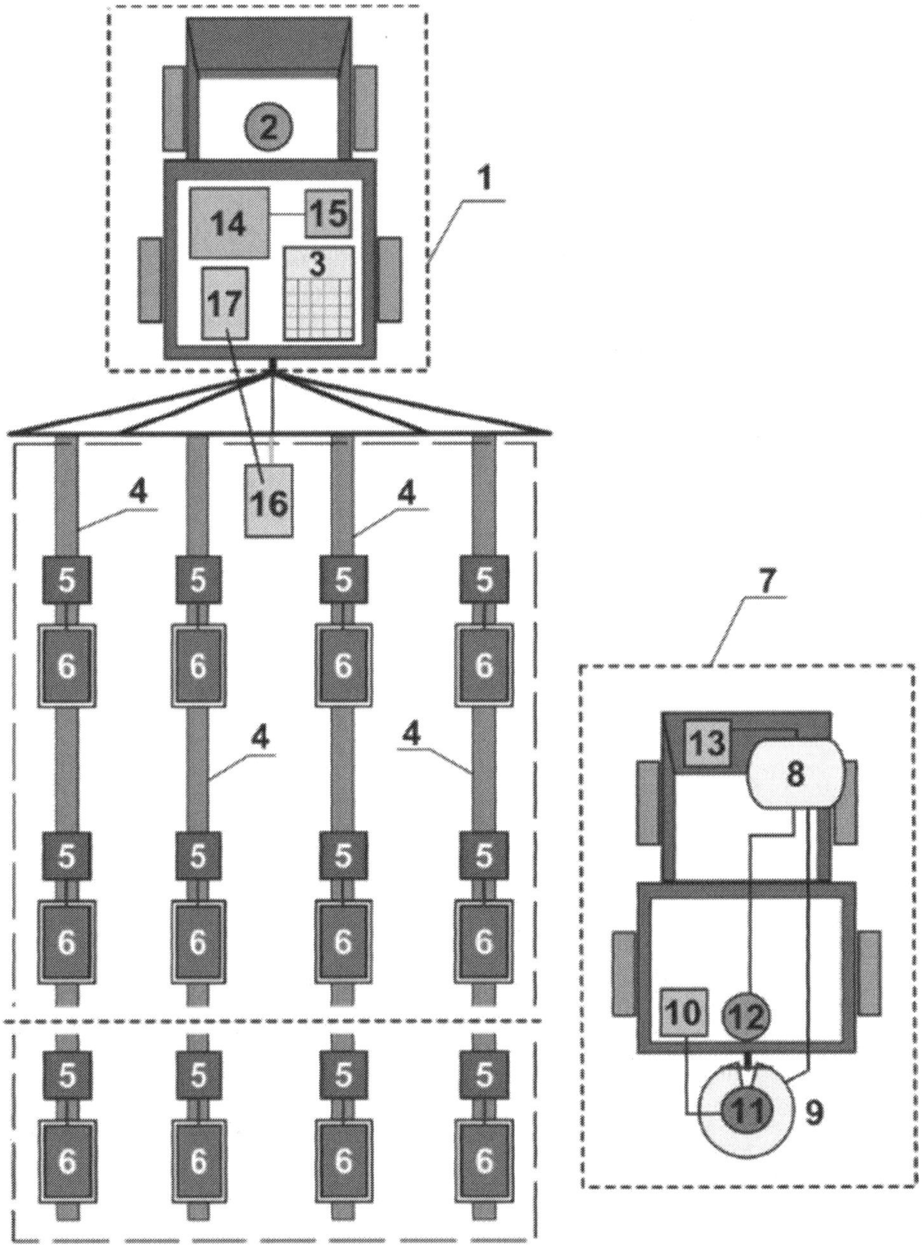

Fig. 1. Basic components of the Seismobile system: 1 – towing vehicle, 2 – GPR receiver, 3 – docking station, 4 – woven belt, 5 – seismic sensor (typically 1C or 3C geophone), 6 – measuring module, 7 - mobile weight drop, 8 – weight drop control unit, 9 – pile driver, 10 – measuring module, 11 – accelerometer, 12 – GPS receiver, 13 – radio modem, 14 – database server, 15 – Wi-fi module, 16 – GPR antenna, 17 – GPR control unit, (on the basis of Isakow et al.[15]).

Table 1. Parameters of seismic acquisition used in tests [8].

Parameter	**Value/description**
Geophone natural frequency	10 Hz
Sampling period	0.125 – 0.250 msec
Seismic record length	500 – 1500 msec
File format	SEGY
Source type	electrodynamic Gisco ESS200 weight drop (90 kg) and hammer impact (5 kg)
Stacking	8
Distance between landstreamer lines	2 – 3.5 m
Geophone intervals	1 – 5 m

3 Testing wireless data acquisition

During Seismobile testing, multichannel seismic recording comparisons were obtained using a configuration of 80 channels (4 landstreamer 24-channel belts, each filled with 20 measuring modules) with different seismic sources, including a traditional 5-kg hammer, a 90-kg Gisco pile driver, and a custom-made 45-kg pile driver fixed to a special mini KAWASAKI technical car. Figure 2 shows the preparation of the Seismobile measuring set for testing, and Figure 3 shows the entire test track.

Fig. 2. Preparation of the measuring set for Seismobile testing (assembly of sensors and measuring modules on the tape of S4 streamer), and excitation of a seismic wave signal using the 90-kg pile-driver (trailer) and the 45-kg pile-driver mounted to the KAWASAKI mini car.

Figure 4 shows examples of seismic records obtained during wireless data acquisition testing of the Seismobile system. Figure 5 shows sample tests with a summation of results, and Figure 6 compares the frequency range (average spectrum of 10 strokes) of the waves from a 5-kg hammer and 45-kg and 90-kg pile drivers at a distance of 1 m from sensor no.11 (the sensor located in the middle of the landstreamer belt).

Fig. 3. Set of four streamers belts, 100 m long, with different spacing (S1-S2 1 m, S2-S3 1.5 m, S3-S4 1 m), each equipped with 20 GS-11D sensors spaced 4 m apart.

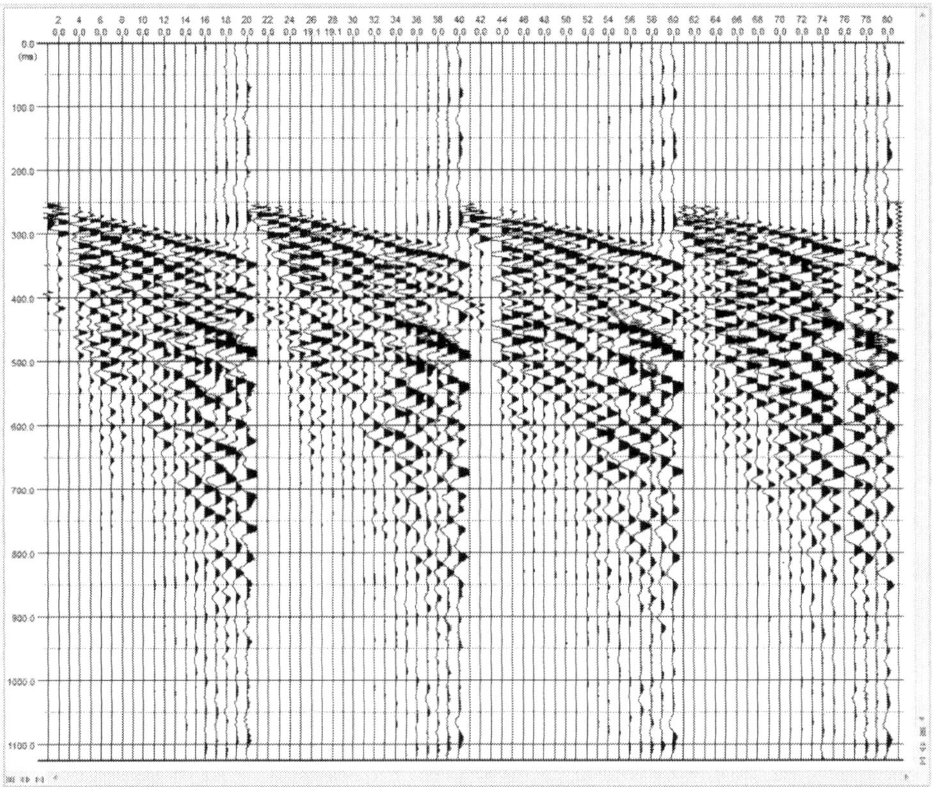

Fig. 4. Examples of wireless seismic recordings registered by the system during testing (45-kg pile-driver source, all sensors on tapes of the streamers, excitation at geophone no. 1).

 DOI: 10.1051/e3sconf/20172401008

During the study, particular attention was paid to the synchronization of source timing, the measurement of geometry by autonomous GPS systems, and the repeatability of triggering measurements constrained by an accelerometer identifying the seismic waveform. The dynamics of system registration, the reliability of wireless transmission, and transmission range were also examined. The automatic numbering of the measurement modules was tested as they were laid out and fixed to the landstreamer belts. At the end of the measurement session, the accuracy and speed of data downloading from the internal memory (SDHC 32GB WiFi) of the measurement modules located in the docking station were checked. The functionality of automatic module battery charging when docked, the operating life after battery charging, and the correct signalling of module discharge state were evaluated, as well.

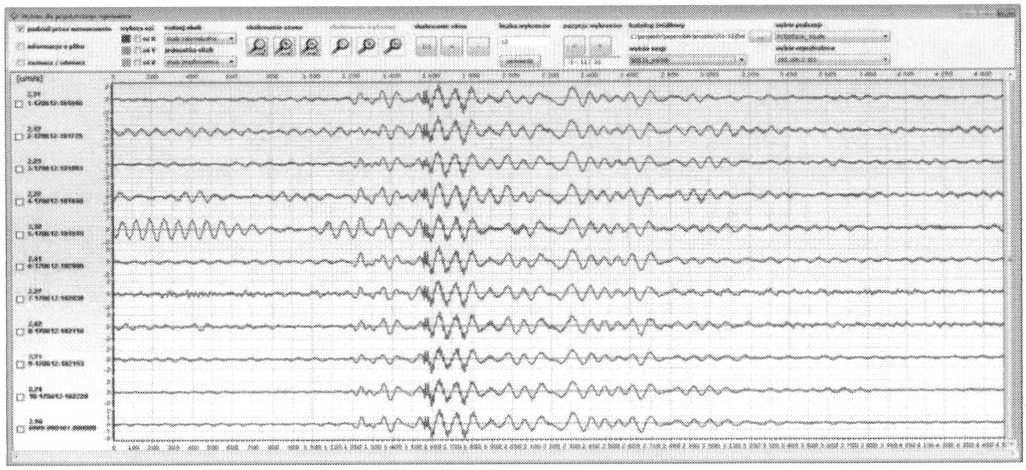

Fig. 5. Example of summing up of 10 recordings excited with a hammer of 5-kg mass for the sensor no. 11. The top 10 graphs show the components of the sum, while the latest waveform shows the averaging result.

4 Conclusions

The Seismobile system tests demonstrate the suitability of the system for field trials. Direct digital recording using internal Flash memory, and the high dynamics of A/D processing provide high quality data, undisturbed by the influence of external magnetic fields. Wireless viewing of seismic records during data acquisition enables a quick assessment of the measured data, while the transmission range is sufficient for measuring at a distance of 100 m. Wireless data logging in Flash memory modules at the end of data acquisition during battery charging ensures reliable archiving of survey data.

The use of uni- and tri-axial sensors ensures the best (in terms of frequency spectrum, amplitude attenuation, and mechanical resonance minimization) transmission of vibrations from the ground. Studies have shown that compared to a 90-kg pile driver, a 45-kg pile driver delivers approximately 30% less energy to the subsurface, and a 5-kg hammer delivers approximately 80% less. Demonstrated repeatability of summation of the recorded components without the "trembling" of the signal phase enables the results to be summed effectively.

The 90-kg pile driver generated the lowest frequencies, and the 45-kg pile driver generated more energy at higher frequencies (primarily due to the use of a steel intermediate plate). Fully charged MP modules can work continuously for at least 8 hours, and the charging time of fully discharged battery modules does not exceed 2 hours.

 DOI: 10.1051/e3sconf/20172401008

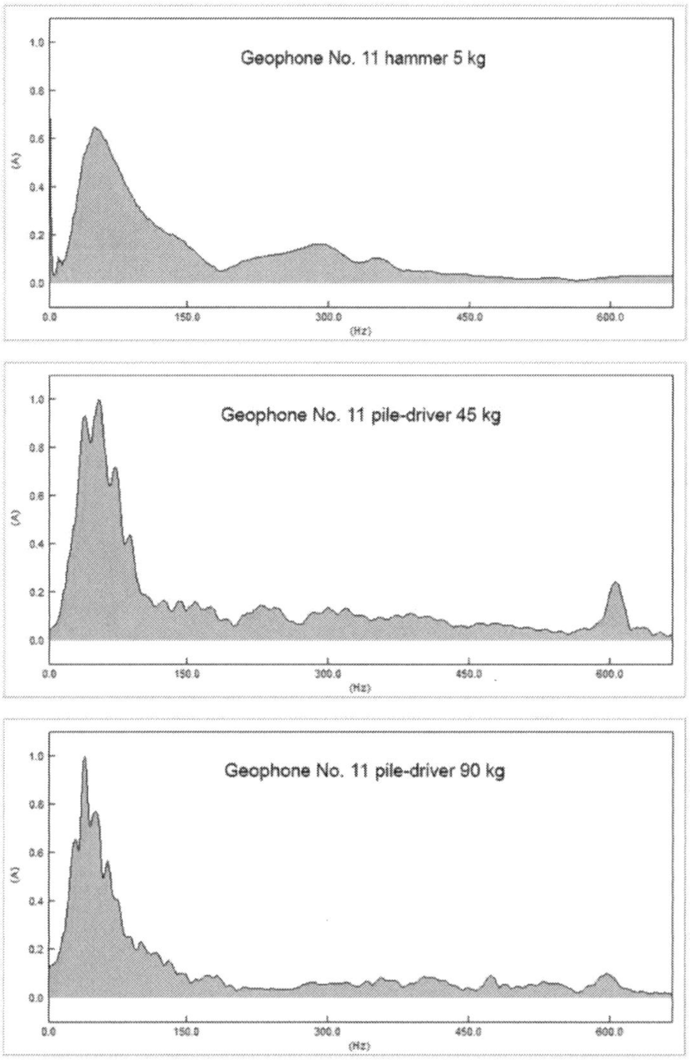

Fig. 6. Comparison of the frequency range (averaged spectrum of 10 strokes) of the excited wave using a 5-kg hammer and the 45-kg and 90-kg pile-drivers at 1 metre distance from sensor no. 11.

The article was prepared as a result of the SEISMOBILE project No. UOD-DEM-1-303/001 performed within the Applied Research Programme DEMONSTRATOR+ co-financed by the Polish National Centre for Research and Development.

References

1. J. Hugenschmidt, J. Appl. Geophys. **43**, 2-4, 147–155, (2000) doi 10.1016/S0926-9851(99)00054-3
2. T. Inazaki, Explor. Geophys. **35**, 1–6 (2004) doi 10.1071/ EG04001
3. R.D. Miller, *High-resolution seismic reflection to identify areas with subsidence potential beneath U.S. 50 highway in Eastern Reno County, Kansas*. Proc. Symposium on the Application of Geophysics to Engineering and Environmental Problems 2006, 460–472 (2006)

4. Z. Pilecki, *The role of geophysical methods in the estimation of sinkhole threat in the post-mining areas of shallow exploitation in the Upper Silesian Coal Basin, Poland* Gospod. Surowcami Min. **24**, 3/1, 27–40 (2008)
5. Z. Pilecki, *Methodology for A-1 motorway basement treatment effectiveness improvement by means of geophysical mrthods in the areas of metal ores shallow mining threatened with the sinkhole occurence in the Upper Silesia*, Gospod. Surowcami Min. **25**, 3, 319–331 (2009)
6. A. Miłkowski, Z. Pilecki, K. Kłosek, M. Tondera, *Autostrada A1 zaprojektowana na „dziurawym" podłożu, Cz. 1*, Magazyn Autostrady: Budownictwo Drogowo-Mostowe **3**, 104–112 (2010)
7. C. Krawczyk, U. Polom, S. Trabs, T. Dahm, J. Appl. Geophys. **78**, 133–143 (2012) dx.doi.org/10.1016/j.jappgeo.2011.02.003
8. Z. Pilecki, Z. Isakow, R. Czarny, E. Pilecka, P. Harba, M. Barnaś, J. Appl. Geophys. **143**, 31–41 (2017) doi 10.1016/j.jappgeo.2017.05.016
9. P. Harba, Z. Pilecki, Landslides **14**, 1225–1233 (2016) doi 10.1007/s10346-016-0779-2
10. Z. Pilecki, J. Kłosiński, E. Pilecka, J. Karczewski, J. Ziętek, *Wpływ zawodnienia osuwiska na rozpoznanie granic jego nieciągłości metodami falowymi*, Gospod. Surowcami. Min. **24**, 2/3, 427–444 (2008)
11. Z. Pilecki, P. Harba, M. Laszczak, A. Adamczyk, S. Cielesta, *Strimery w sejsmice inżynierskiej*, Przegląd Górniczy **7**, 32–38 (2014)
12. A. Kubańska, Z. Isakow, Z. Pilecki, *Założenia funkcjonalne systemu Seismobile*, Zeszyty Naukowe - Bulletin of the Mineral and Energy Economy Research Institute of the Polish Academy of Sciences **93**, 133–142 (2016)
13. Z. Pilecki, Z. Isakow, A. Kubańska (eds.), *Seismobile – system geofizycznego badania podłoża szlaków komunikacyjnych* (Bater Drukarnia MY-HORYZONT, Katowice, 2016)
14. Z. Isakow, K. Siciński, P. Sierodzki, *Innovative solutions for data acquisition in Seismobile landstreamer*, Zeszyty Naukowe - Bulletin of the Mineral and Energy Economy Research Institute of the Polish Academy of Sciences **93**, 155–168 (in Polish) (2016)
15. Z. Pilecki, R. Czarny, M. Chamarczuk, K. Krawiec, E. Pilecka, *Skuteczność rejestracji zestawu pomiarowego strimera sejsmicznego Seismobile*, Zeszyty Naukowe - Bulletin of the Mineral and Energy Economy Research Institute of the Polish Academy of Sciences **93**, 143–154 (in Polish) (2016)
16. A. Malehmir, B. Brodic, M. Dehghannejad, C. Juhlin, E. Lundberg, *A state-of the-art mems-based 3c seismic landstreamer for various near-surface applications*, Proc. 78th EAGE Conference & Exhibition – Workshop Programme, Vienna, 20 May – 2 June 2016, WS04 B02 (2016)

Causes of some hazardous engineering-geological processes on urban territories

Tetiana Kril[1*]

[1]Institute of Geological Sciences of NAS of Ukraine, Engineering-Geological Department, 01054 Kyiv, Ukraine

Abstract. Population growth in cities, the need to expand the living space requires of rational use of territories within the existing boundaries of the city. The necessity of compliance with the functional zones of the city is shown on the example of a representative part of Kiev, that should be performed taking into account engineering-geological features of the territory. It is necessary to comply with the underlying zones in the underground space to ensure the bearing capacity of the soil mass. The changes in soil bases are defined as a result of changes in the stress-strain state under the construction, development of underground space, changes of soils water content as the result of soaking from the surface, formation of "perched water", raising the groundwater level. The vibration analysis of high-rise building – the main library building is made from the dynamic loads that arise during the movement of the vehicle, taking into account the work of the pile foundation as a rigid body relative to the longitudinal axis, which passes through the center of the building at the level of the cap of piles.

1 Introduction

The territory of any city can be divided into four main functional zones: residential, industrial, transport, recreational. The formation and placement of the objects is regulated by the planning documentation [1] and the Law of Ukraine "On regulation of urban development" valid in Ukraine. From the point of view of the engineering development of the territory for each type of functional zones according to natural conditions there are three categories: favorable, slightly favorable and unfavorable [2].

Each of the functional zones is characterized by a different degree of anthropogenic influence, ensuring the safety of residence / stay for a person and the mutual influence of objects from one zone to another (particularly in the underground semi- space).

The residential area is the location of buildings and constructions of permanent human habitation. The main influence on the geological environment in such zones is caused by a static load, which varies from 63 kN/m^2 (5-storey building of 1960-ies) to 380 kN/m^2 (modern high-rise buildings of 27 storeys). In industrial zones there are powerful industrial objects and power plants, which creating static and dynamic additional impacts on the soil,

* Corresponding author: kril@nas.gov.ua

and also in case of accidents or improper operation of significant leaks from water supply networks.

The vibrations acceleration changes in the range 30-130 dB (0,05-1,8 in g) [3], that is transmitted to the soil from the movement of different types of transport in the transport infrastructure zones. Recreation zones such as parks and other recreational facilities, do not create significant negative influences. Generally for these zones within the city use areas with not the best engineering-geological conditions.

Large cities are characterized by non-compliance of functional zones, for example, energy objects are not systematically located, sometimes along with residential buildings; the road of categories with high traffic loads are close to residential neighborhoods. The close mutual arrangement of objects together with complex engineering-geological conditions can lead to an increase in the intensity of such influence and generation of hazardous engineering-geological processes.

The differentiation of the territory of Kiev in terms of the vulnerability of its geological environment to man-made influences was carried out. The method of quadratic evaluation was used, with the dimensions of a unit square of 300×300 m [3]. Factors such as static load, dynamic load, type of soil with load bearing capacity, ground water level etc., were taken into account. Squares with unfavorable engineering-geological conditions (60 % of the unit squares) have an intense mixing of functional zones in most cases.

The aim of this article was to show the occurrence of hazardous engineering-geological processes as a result of mutual influence of objects belonging to different functional zones, and to describe the conditions of violation of the geological environment under the urban planning activities, on the example of a high-rise building located on one of these sites.

2 The site description and methodology

The experimental site with a high-rise building – the building of the Vernadsky National Library of Ukraine is placed in the Holosiivskyi district of Kyiv and occupies 0.18 km^2. The objects of different functional zones are bordered on this territory: on the two sides at a distance of 50-80 m the building is surrounded by 12-16-storey residential buildings; within a radius of 60-100 m from different directions – roads with high transport load, including a traffic interchange and railway tracks; there are three food processing plants and a petroleum – petrochemical plant (Fig. 1a).

The site is located in the lower part of the right bank slope of the river valley Lybid within the Prydniprovs'ka upland, by geomorphologically. The absolute marks of the surface of the site are 115.5-120.7 m. The library complex consists of the high-rise 27-storey bookstore on a pile foundation and a 3-, 4-storey buildings on a strip foundation, set around the high-rise part (stylobate). The stylobate is separated from the high-rise book storage by expansion joints.

In 2008-2010, during the reconstruction of the traffic junction the Moscovs'ka Square and the construction of the subway, there were recorded the appearance of cracks on the constructions of the library building and damaging the covering on the facade (up to 20 % of the total area) that is associated with such building activities (Fig. 1b).

The documentations of engineering-geological exploration, a survey of the technical condition of the building, the simulation of the stress-strain state of the soil basis, the impact of dynamic loads, changes in water level were processed in order to determine the causes of the damages. Statistical, cartographic methods were used to build of all the charts using GIS technology and digital modeling. The mechanics and mathematical bases of engineering geology in combination with the approaches of system analysis and engineering-geological similarity theory were used.

 DOI: 10.1051/e3sconf/20172401009

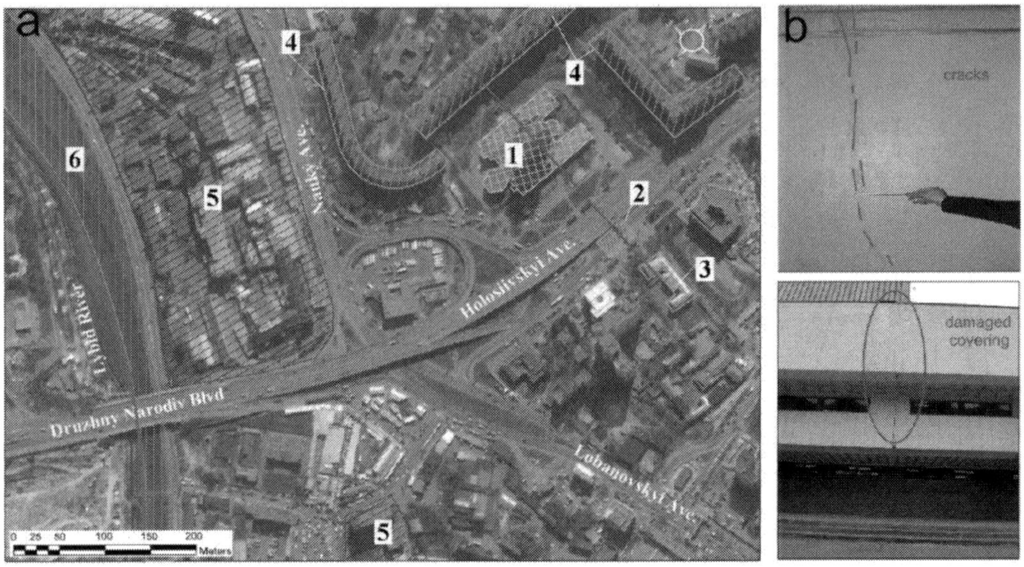

Fig. 1. The plan of the experimental site (a) and the damage on the constructions of building (b): 1 – the building of the library; 2 – the metro station; 3 – the school; 4 – residential buildings; 5 – industrial enterprises; 6 – railway tracks.

3 Results and analysis

3.1 The change of the stress state under the building and planning of underground space

The influence on the geological environment during urban planning is the stages of destruction of the natural environment (destruction of plant and soil cover, digging pits and excavation), the inclusion of new elements into the natural system (construction of buildings, underground structures, coverage areas, etc.).

Ensuring the stability of residential buildings, especially with the increasing number of floors, at the same time requires an increase in the depth of the foundation, the replacement of natural bases with artificial, which significantly changes the magnitude of stress in the first surface components of the geological environment. The general scheme of changing the stress state of the geological environment is shown in Fig. 2 on the example of the high-rise book storage.

Engineering constructions most actively affect on the soil layers close to the surface. The weight of the construction forms a compressed zone under the foundations at a depth of 10 to 100 m, respectively, and in lateral directions in the corners of the building up to 30 m.

The values of the compressed zones under the foundations of buildings along the line of the section I-I were established by the calculations, that is from 10 m (3-storey school building) to 50 m (27-storey library), according to building codes [4].

High-rise buildings with underground parking that built on piled foundations violate the hydrogeological regime of the first from the surface aquifer and create a barrage effect. The rise of the water level to the deep security zone or the zone of active compression under the existing objects generates the subsidence process in the loess types of soil. Similar conditions have developed on the territory of the reserve of world significance "Sophia of Kiev" during the construction of a high-rise building in its buffer zone [5].

Under the vertical protective zone of the building should be understood a zone around the bulb of the active compressed zone under it. Its size depends on the weight of the building

DOI: 10.1051/e3sconf/20172401009

and the depth of laying the foundations for different types of soils, and the intervention in which can lead to emergency situations.

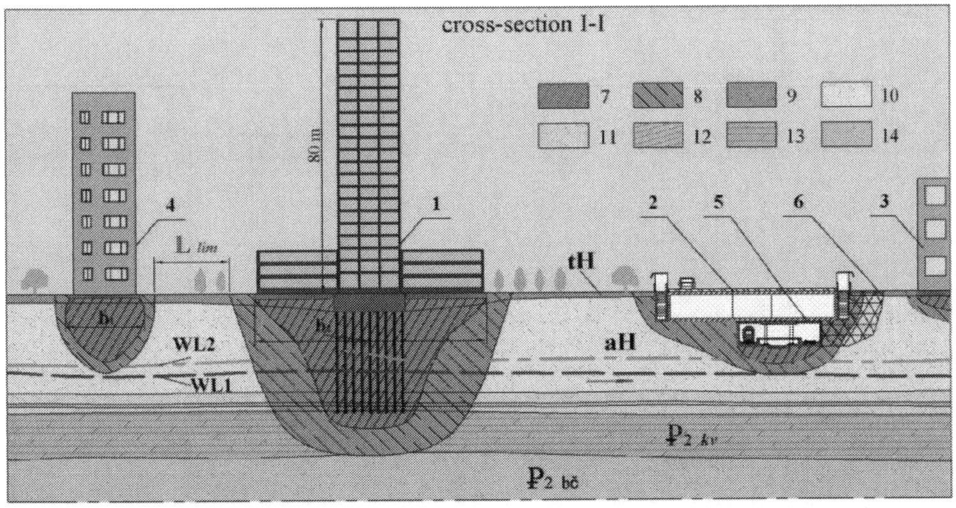

Fig. 2. The scheme of changing the stress field of the geological environment under the main objects of urban development along the line of the section I-I: 1 – the construction of the library; 2 – underground passage; 3 – the school; 4 – the residential building; 5 – the metro station; 6 – the place of the accident in 2006; 7 – a zone of active pressure under construction; 8 – the security zone; 9 – filling soils; 10 – loamy sand of loess-like type; 11 – fine sands; 12 – loam with organic matter; 13 – marl clay; 14 – dense sands; WL1 – water level; WL2 – changed water level due to the barrage effect; L_{lim} – the limit between protective areas of two buildings; b_1, b_2 – the maximal width of the compressed zone under the foundation; tH – filled soils, aH – alluvial soils, sands and loamy sands, $₽_{2\,kv}$ – clay marl of kyyivska suite, $₽_{2\,bc}$ – sandy soils of buchatska suite.

The distance from new buildings to existing buildings is a major parameter. A reduction in the distance threatens to changes in geological environment and leads to damage or destruction of the buildings. Therefore, the question of the limit distance (L_{lim}) between the security zone and the active compressed zones of new objects is relevant for the build at the intersection of various functional zones of the city (Fig. 2).

The appearance the barrage effect is most likely at a perpendicular location of long axis of new buildings with the direction of unloading of underground water. The value L_{lim} should be calculated taking into account the geometric parameters of the foundation and to consider depth of groundwater:

$$L_{lim}=(b_1+b_2)/2+W, \tag{1}$$

b_1, b_2 – the width of the compressed zone under the foundations of neighboring buildings, W – the distance at which the barrage effect is minimized.

3.2 Changes in soil water content

The territory of the experimental site refers to the potentially flooded by waters of the "perched water" and the main aquifer. There are two water levels. The first aquifer of the type "perched water" lies at a depth of 2.1 to 4.0 m, this layer is low-power. Its formation is associated with atmospheric precipitation, which are trapped in the relatively impermeable lenses of the loamy soil. The second aquifer formed in deluvial loam, sandy loam and alluvial sands is more powerful. The confining layer of this aquifer is the marl clay of the Kievan suite of the Paleogene.

Sandy-clay soils are the bases of all structures in this area, and change the bearing capacity depending on the soil water content. The changes in the strength parameters of soils depending on their humidity are shown in the diagram (Fig. 3).

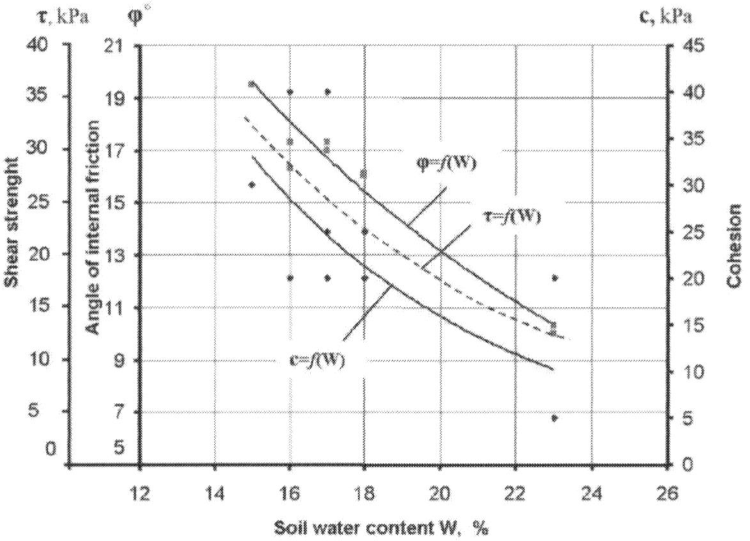

Fig. 3. The graph of the dependence of the shear strength, the angle of internal friction and the cohesion of the quaternary loamy sands from the water content.

In the modeling of the situation of saturation of the soil bases through raising the level of groundwater by 10 m (the appearance of the "perched water"), using the Coulomb-Mohr model, it was established that the value of the active pore pressure increased by 44 % (to - 398.77 kPa), the effective stresses decreased by 16% (to 872.52 kPa), which led to an increase in deformations under the foundations of buildings up to 62 % (total displacement $182.45 \cdot 10^{-3}$ m). The modeling showed that the greatest influence on the formation of plastic zones and the development of processes of subsidence occurs under the construction of shallow laying, in particular the metro station.

The decrease of the strength as a result of additional moisture can lead to emergency situations [5, 6]. During the digging of the pit for construction of the metro station "Demiivs'ka" in January 2006 an accident occurred: wall which protected the foundation from the side of the school suddenly collapsed, and there was an ejection of the soil into the pit. The accident location is shown in the Fig. 2, (5). There was a failure of a number of communications and the traffic was blocked for a few days. The causes of the accident include the coincidence of anthropogenic and natural factors: as a result of the rising water content of the loamy sand by leakage from the water supply and further expansion of the soil after frost.

3.3 The influence of the traffic induced vibrations

The library complex with the high-rise building is located in the zone of influence of several dynamic sources: the traffic junction the Moscovs'ka Square; Holosiivskyi Ave. with heavy traffic (average value 3.8 thousand vehicles per hour) Nauky Ave. (average value 4.2 thousand vehicles per hour) and the metro station "Demiivs'ka".

The vibration level from automobile and rail transport generated in the range of 10-120 dB [3], which initiates the displacement of soil particles at high speeds (from 6 to 200 μm/s). The level of the oscillations in a geological environment from such number of sources can be compared to weak earthquakes in the 1-2 level in the MSK-64 scale. The level of vibration from metro stations of shallow laying exceeds 4 times the seismic noise [7].

On the seismic noise, which is the result of either remote areas of earthquakes, or geodynamic activity of that area, vibrations from these sources are distributed in the soil environment, mainly in the upper part of a thickness up to 10-15 m, transmitted to the foundations and building construction of the library and other objects.

The basis of the foundations is the thickness of the loamy sand of loess-like type from 9 to 16 m. The velocity of propagation of seismic waves in loamy sand is 220 m/s, and at the additional moistening is increased to 275 m/s. If the water level rises to 4 or to 2 m, the seismic intensity will increase on 0.6 or 0.8 of the level in the MSK-64 scale, respectively.

If the oscillation frequencies from one or more sources come in resonance with the natural frequencies of the soil bases or foundations of the building, the amplitudes of the displacement can increase in 10-12 times [3], which may be affected by the additional expansion of the cracks or the increase of their number in the building structures.

Since the building is constructed in parallel with the axis of tunnel of the subway, therefore, its lowest resistance is in the plane normal to the longitudinal axis.

The vibration analysis of the high-rise building under the action of seismic waves based on the work of the pile foundation as a rigid body held relative to the longitudinal axis passing through the center of the building at the level of the cap. To determine contact stresses and forces acting on the construction from the piles, the soil is taken as the Winkler model to simplify the calculations. Dynamic impact from the metro station were determined by the parameters of the soil vibrations. Taking into account the shallow laying of the metro, it is possible to accept, that the maximal effect of the oscillation waves on piles is occur in the horizontal direction.

The vibration equation of the beam as the beam of vertical position under the action of horizontal wave is:

$$\frac{EI}{l^4}\frac{\partial^4 v}{\partial x^4} + \mu\frac{\partial^2 v}{\partial t^2} + vC_zD = C_zD(u_1 + u_2) \tag{2}$$

E – a module of elasticity of the beam material, MPa; I – a moment of inertia of the cross-section of the beam, m^4; l – the insertion length of the pile, m; μ – the linear mass of piles, kg/m; υ – the function of moving the beam; u_1, u_2 – the moving in the coming and reflected waves, m.

$$u_1 = u_2 = u_0 \cdot sin\omega t \tag{3}$$

ω – circular frequency of the coming seismic waves, s^{-1}; D – the diameter of the pile, m; Cz – the compression ratio for the Winkler model, kN/m^3.

The calculation of action on the building from piles, enclosed in the cap was done, not accepting to the calculation the horizontal moving of the cap in relation to piles at such boundary conditions:

$$x = 0 \quad \frac{\partial^3 v}{\partial x^3} = \frac{\partial^2 v}{\partial x^2} = 0 \tag{4}$$

$$x = l \quad v = \frac{\partial v}{\partial x} = 0 \tag{5}$$

The coefficients of the normal forms are determined by formulas:

$$a_i = \frac{2C_zDu_0\int_0^1 X_i dx}{\mu(p_i^2 - \omega^2)}, \; p_i^2 = \frac{C_zD + k_i^4\frac{EI}{l^4}}{\mu} \tag{6}$$

$$\int_0^1 X_i dx = \int_0^1 \left(V(k_ix) - \frac{T(k_i)}{S(k_i)}U(k_ix)\right) \cdot dx = \frac{chk_i \cdot cosk_i - chk_i - cosk_i}{2K_iS(k_i)} \tag{7}$$

the value of k_i^2 is given in the Table 1 (for four forms of oscillations), S, T, U, V – the function of N. Krylov [8].

Table 1. The coefficients of the *i*-th natural frequency.

i	1	2	3	4
k_i^2	3,516	22,03	61,67	120,9

For high-rise part of the building:

$$E = 3 \cdot 10^{10}\ Pa;\ l = 17{,}8\ m$$

$$I = \frac{\pi}{4} r^4;\ \ r = 0{,}4\ m;\ \ I = 200{,}96 \cdot 10^{-4}\ m^{-4}$$

$$\mu = 2{,}7 \frac{\pi D^2}{4} 10^3 = 1{,}36 \cdot 10^3\ kg \cdot m^{-1}$$

$$p_1^2 = 2{,}4 \cdot 10^4\ s^{-2};\ \ p_1 = 119{,}3\ s^{-1};\ \ f_1 = 19\ Hz$$

$$p_1^2 = 2{,}7 \cdot 10^4\ s^{-2};\ \ p_1 = 140{,}7\ s^{-1};\ \ f_1 = 22\ Hz$$

$$p_1^2 = 3{,}8 \cdot 10^4\ s^{-2};\ \ p_1 = 242{,}4\ s^{-1};\ \ f_1 = 38\ Hz$$

$$p_1^2 = 10{,}7 \cdot 10^4\ s^{-2};\ \ p_1 = 431{,}08\ s^{-1};\ \ f_1 = 68\ Hz$$

The reaction to the action of piles in the cap:

$$P_i = \frac{EI}{l^3} \cdot a_i X_i''(0) \tag{8}$$

$$P = \sum_{i=1}^{3} P_i = 1{,}5 + 4{,}7 + 61{,}3 = 67{,}5\ N$$

The moment acting from the piles on the cap:

$$M = \frac{EI}{l^2} \sum_{i=1}^{3} a_i X_i''(0) \tag{9}$$

$$M = \frac{3 \cdot 10^3 \cdot 200{,}96 \cdot 10^{-4}}{17{,}8^2} \cdot \left(0{,}49 \cdot 10^{-6} \cdot 8{,}45 + 0{,}27 \cdot 10^{-6} \cdot 54 + 0{,}53 \cdot 10^{-6} \cdot 147{,}5\right) = 184{,}4\ \text{Nm}$$

If this calculation continues for the option of free bearing of the cap on the pile, the effect of the horizontal force will decrease, and the bending moment is eliminated. However, in this case, since the building is high, the condition of enclosed of the piles into the cap is necessary.

Reduce of the influence of lateral forces depends on the ratio of the frequencies of piles and frequency of oscillations, which are excited by waves. It is possible only with a special arrangement of piles in any ratio between own and forced frequency. The piles are perpendicular to the front of the propagation waves and have distance between them equal integer number of waves.

4 Conclusions

Violation of the conditions of safe habitation in cities is connected with the unsuccessful placement of objects of different functional zones that have different degrees of safety of the inhabitation of people. It happens in case of hazardous engineering-geological processes, together with emergency situations. Changing the water level and change in soil water content have the most dangerous influence among the considered effects. The effect of soil water content is indirect and showed itself in the following:

– the increase of pore pressure at raising the water level leads to changes in the redistribution of stresses in the underground space, and the increase of vertical displacements of soil particles, decreases the maximum distance between buildings;

– the increase in soil water content leads to an increase in the velocity of propagation of longitudinal waves, therefore, in percolated soils, the distance distribution of the dynamic load increases and can reach the object, the distance to which from sources does not corresponds to the standards.

In built-up areas, leakages from engineering networks (water supply) that occur due to an emergency state of communications, and excessive atmospheric precipitation, which depend on the condition of storm drains and drainage structures are main sources that lead to changes in soil water content.

The close location of high-rise buildings on piles to residential buildings on shallow (tape or slab) foundations is hazardous in the case of high occurrence of the first aquifers. The raising of the groundwater level due to the barrage effect through the dense arrangement of piles can lead to uneven settlement and the buildings roll of residential buildings in the direction of high-rise buildings.

The comfort of staying in a building with a close location of transport routes depends on the parameters of the damping of the type of soil, its physical condition and the structural design of the foundations. The calculation showed when the free supporting the piles on the caps occurs less impact for a modern high-rise buildings.

Today, legislation regulates zoning of settlements only on the surface (horizontal). The widespread intervention into the underground space (on the territory of Kiev of the 1970-ies has built more than 1000 high-rise buildings with an underground depth from 25 to 80 m).

The influence of the underground part of a new or reconstructed engineering object should not interfere with the deepening zone of other existing building. Therefore, the development of standards for deep (vertical) zoning become relevant, that should be done depending on the geological features of the built-up area and surrounding buildings.

The functional zones of the city should be developed taking into account not only the social and economic-planning factors, but also to compare it with the engineering-geological conditions, the geodynamic features of the territory (tectonic disturbances, changes in the lithological composition of the soils, etc.), anthropogenic load of the territory (building, etc.), that is connected with the safety of people.

References

1. *Urban planning. The planning and development of urban and rural settlements. DBN360-92. Valid from 2007-01-02* (Ministry of Regional Development of Ukraine, Kyiv 84, in Ukrainian, 2006)
2. M. Demchyshyn, T Kril, *Geodynamic conditions as the main factor of functional zoning of settlements. Modern information technology management environmental safety, environmental management, measures in emergency situations*, Proc. of the 11th Int. Sci.-Practic. Conf., Kyiv – Kharkiv – Crimea, 281-288 (in Ukrainian) (2012)
3. T. Kril, *Technogenic dynamic influences on the geological environment of city (on an example of Kyiv)*, Naukova dumka, Kiev **160** (in Ukrainian) (2015)
4. *Bases and foundations of buildings, General design: DBN.2.1-10-2009. Valid from 2009-07-01* (Ministry of Regional Development of Ukraine, Kyiv 104, in Ukrainian, 2009)
5. M. Demchyshyn, T Kril, O. Anatsky, N. Sokovnina, S. Vdovichenko, *Modern information technology management environmental safety, environmental management,*

measures in emergency situations, Proc. of the 15-th Int. Sci.-Practic. Conf., Kyiv, 142-148 (in Ukrainian) (2016)

6. M. Demchyshyn, T Kril, O. Anatsky, *Modern information technology management environmental safety, environmental management, measures in emergency situations*, Proc. of the 15-th Int. Sci.-Practic. Conf., Kyiv, 148-152 (in Ukrainian) (2016)
7. S. Shcherbina, T. Kril, *Information system for analysis of temporal variation of ground accelerations generated by metro trains*, Modern Science **3**, 30-36 (2014)
8. V. Biderman, *Theory of mechanical oscillations* (Vyshaja shkola, Moscow, in Russian, 1980)

E3S Web of Conferences **24**, 01010 (2017) DOI: 10.1051/e3sconf/20172401010
AG 2017 - 3rd International Conference on Applied Geophysics

The impact of the structural features of the rock mass on seismicity in Polish coal mines

Renata Patyńska[1*]

[1]Central Mining Institute, 40-166 Katowice, Poland

Abstract. The article presents seismic activity induced in the coal mines of the Upper Silesian Coal Basin (GZW) in relation to the locations of the occurrence of rockbursts. The comparison of these measurements with the structural features of the rock mass of coal mines indicates the possibility of estimating the so-called Unitary Energy Expenditure (UEE) in a specific time. The obtained values of UEE were compared with the distribution of seismic activity in GZW mines. The level of seismic activity in the analysed period changed and depended on the intensity of mining works and diverse mining and geological conditions. Five regions, where tremors occurred (Bytom Trough, Main Saddle, Main Trough, Kazimierz Trough, and Jejkowice and Chwałowice Trough) which belong to various structural units of the Upper Silesia were analyzed. It was found out that rock bursts were recorded only in three regions: Main Saddle, Bytom Trough, and Jejkowice and Chwałowice Trough.

1 Introduction

The key problem in the mining science and practice is the progress in identifying the direct causes of the occurrence of a rockburst, especially in the phase of the occurrence of this dynamic impulse due to a seismic tremor [1-11]. Then, more effective preventive measures against rockbursts can be performed [12]. Seismic events and related seismic hazards in a given field are determined by a number of natural and technical factors [4].

The detailed knowledge of the structural features of the rock mass of coal mines of the Upper Silesian Coal Basin (GZW) should be a necessary factor for proper assessment of seismic and rock burst hazards and proper design of exploitation [5, 13].

It is becoming clear that, by proper location of the operation in relation to the direction of the fracturing of the rock mass, seismic hazards can be affected and reduced in the entire cycle of exploitation work [12]. It is introduced by the Unitary Energy Expenditure (JWE, J/t), defined as the ratio of seismicity induced by the operation of mines in a given year [14].

* Corresponding author: rpatynska@gig.eu

2 Seismicity of structural units of the Upper Silesian Coal Basin

The analysis of seismic activity distribution in the years 2007-2016 for individual GZW structural units indicates that the tremors are generated by mining operations and are located in five regions: Bytom Trough (NB), Main Saddle (SG), Main Trough (NG), Kazimierz Trough (NK), and Jejkowice and Chwałowice Troughs (NJ) [15-18]. Data obtained in the location of the rockbursts indicate that these events occurred only in three regions of the GZW:

- 11 rockbursts - Main Saddle (SG), including the following mines: Wujek, Śląsk, Bielszowice, Pokój, Staszic, Halemba – Wirek, Mysłowice -Wesoła,
- 2 rockbursts - Bytom Trough (NB), including the following mines: Piekary, Bobrek - Centrum,
- 8 rockbursts - Jejkowice and Chwałowice Troughs (NJ), including the following mines: Rydułtowy - Anna, JAS – MOS, Borynia- Zofiówka -Jastrzębie and Marcel.

The largest number of tremors (11) was recorded during the exploitation of seams deposited in the north and central wing of Main Saddle (SG). The tremors occurring during mining works in the northern and central wings of the Upper Silesian Coal Basin are mainly related to exploitation of the saddle group, which are covered by thick and strong rock layers [7].

The total number of 507 rockbursts recorded in the years 1970-2016 [16], as many as 470 cases occurred in the 500 group seams, including 194 seismic events in the 510 seam.

The data show that of 21 rockbursts (in the last 10 years), up to 13 cases occurred in the seams of the saddle (500), including 6 in the 510 seam. The remaining rockbursts were related to the seams of group 400, 3 seams of group 700.

Referring the above-mentioned data to the rockbursts recorded in 2016, it is possible to say that there is a trend in the above-mentioned statistics. The effects of rockbursts that occurred in 2016, concerned the seam 507.

Analysis of rockbursts regions shows that these phenomena occur during the exploitation of deposits located at a considerable depth among compact rocks [4, 7].

With the increase in the depth of the longwall operation and the preparatory work, there is a systematic deterioration of geological and mining conditions. In connection with the drawing process of coal resources, the mines more and more often decide to conduct exploitation of deposits in areas with faults and/or in the area of former exploitation [6].

Among the reasons for deciding on the risk of rockbursts in the vicinity of mining works, the most frequently mentioned are:

- increase the depth of exploitation of coal seams,
- high degree of exploitation of the deposit,
- exploitation in residual areas of deposits,
- exploitation of local faults (possible to cross the front of mining works) and of a regional character,
- uneven selection of coal seams in the border areas of mines or between mining areas within one mine.

A detailed analysis of mining and geological conditions of the occurrence of rockbursts recorder between 2007-2016 has indicated that there is a link between the induced seismicity and the location of mining excavations in a particular GZW structural unit [15].

The importance of the formation of the equilibrium of the rock mass in the vicinity of the excavation is given in the geological form of the exploited bed, the properties of the rocks forming the bed, as well as - to a decisive degree - the susceptibility to fracture and rockiness of the rocks lying in the roof.

Research on the impact of faults and planes of reduced cohesion has revealed their association with seismic and rockburst hazards [15-17].

These determinations were made based on the analysis of rock bursts in the GZW coal mines in 2001-2015. The reported cases of rockbursts were, in particular (about 95%) of the impact of the fault zones closest to the rockbursts, mainly (about 67%) in local fault zones with discharge up to a dozen meters (<12 m). Only in 2 cases for 42 analysed rockbursts, there was no presence of a fault zone in the area impacted by the event.

In case of 43% of recorded rockbursts, the effects impacted the areas where works fronts were approaching fault zones. Only 10% of rockbursts were recorded at the fault and 5% occured in the case of transitional faults with mining fronts.

It is clear from the data of the rock masses in the vicinity of troughs that 55% of the analysed cases showed effects in excavations at a distance of up to 100 m from the plane of the nearest fault. At a distance of 120 to 200 m, there were only 22% rockbursts and 64% events over 220 m.

Results of rock analysis obtained in specific geological and mining conditions, confirmed by the results of analytical tests, may be of general value only if they are consistent with the broadly understood mining activity. Therefore, for the verification of literary views, it is important to quantify the effects of natural discontinuities on the occurrence of rockbursts.

3 Unitary Energy Expenditure in the conditions of mines threatened by rockbursts

Analysis of Unitary Energy Expenditures (JWE, J/t) under the conditions of recorded seismic and troughs occurring in the GZW in the years 2007-2016, was based on archival rock data and digital records of seismic activity. Data on mining and recorded rockbursts [16-17] shows that, despite declining coal production in the last 10 years, the energy expenditure (JWE J/ton) has been on the upward trend by 2015 (Fig. 1).

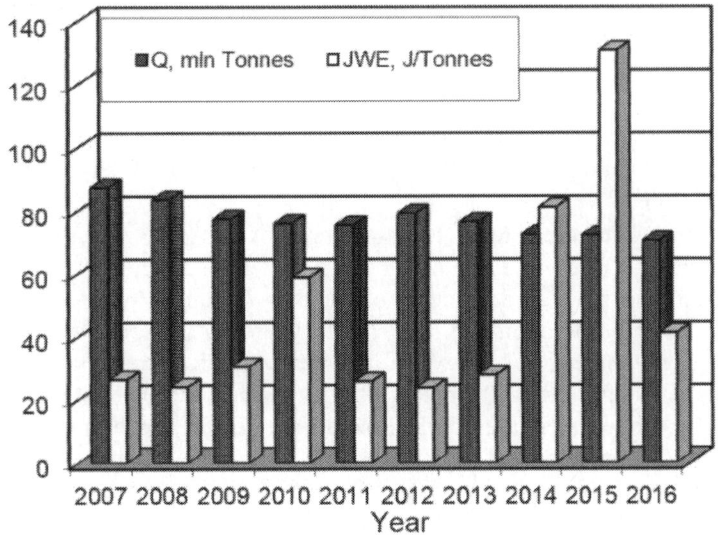

Fig. 1. Unitary Energy Expenditure compared to coal extraction in hard coal mines of GZW in 2007-2016.

Generated seismic activity is undoubtedly related to residual exploitation and, in particular, the concentration of mining and extraction works in the areas of impact residues and exploitation edges. This is evidenced by the fact that, with a steady downward trend in production from about 87.40 million tonnes in 2007 to 70.4 million tonnes in 2016 and a similar number of troughs in the same period: 1-5 a year (Fig. 2), the total seismic activity in

2016, approximately 2.90 $\cdot 10^9$ J, was approximately 18% lower than the average value characterizing the years 2007-2016, 3.53 $\cdot 10^9$ J.

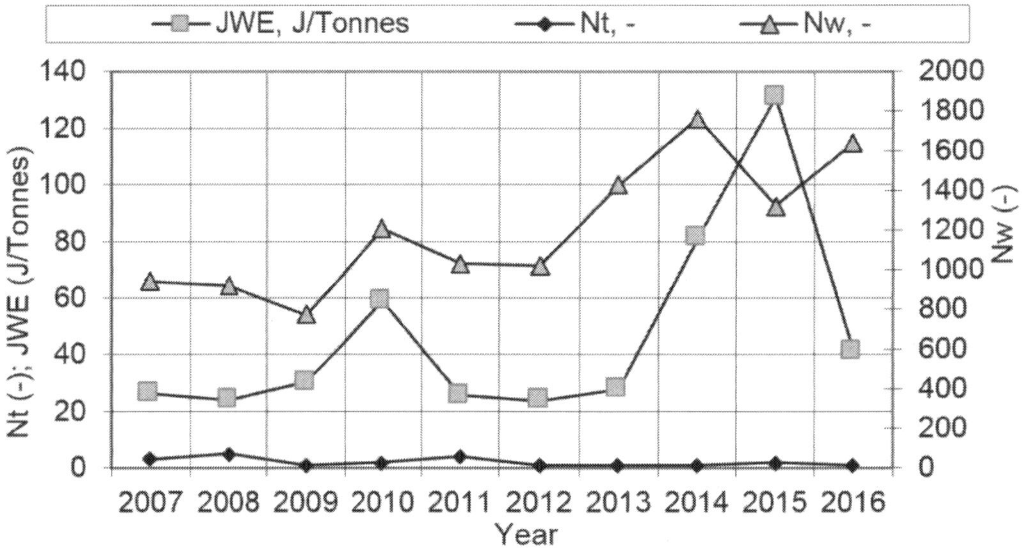

Fig. 2. Unitary Energy Expenditure compared to the number of rockbursts (Nt, -) and number of tremors (Nw, -) recorded in 2007-2016 in the GZW.

Below in Table 1, there are 21 treads against the size of the extraction, with extraction from areas that are at risk of rockbursts. It results from the fact that in the last 10 years total output has decreased by about 19.5%, yet rock mining has increased by about 11%. Half of the mines' output comes from areas subject to rockbursts and remains at 44-53% of total output. Despite a decline in total output in 2016, rock-bottom exploration increased slightly, but increased in particular from seams falling to grade III, which now accounts for about 12% of total output [16].

In order to demonstrate the relationship of the energy of rock mass to the mining operation of a given structural unit of the GZW rock mass, the Unitary Energy Expenditure was estimated for each area.

In total, in 2007-2016 there were 12,044 energy events of $E \geq 10^5$ J, including 9,593 energy phenomena of 10^5 J, and 2,196 of 10^6 J. There were 228 energy events of 10^7 J, and 23 of 10^8 J. At that time, the recorded energy of 4 events amounted to 10^9 J (Stec 2008-2017).

Although the mining operation over the past 10 years has covered the entire GZW area, rockbursts have been concentrated in five of its areas belonging to different structural units. The hypocentres show that the rock bursts have occurred in those areas where the deposits of coal are deep, and the layers are characterized by a high proportion of sandstone layers and strongly developed tectonics. They are: Main Saddle (SG), Bytom Trough (NB), Main Trough (NG), and Jejkowice and Chwałowice Trough, and Kazimierz Trough (NK.

Seismic activity in areas of mines where currently mining exploitation is not carried out, does not occur. On the other hand, recorded seismic activity in active mines, as the conditions of mining works deteriorate, increased. The most frequently mentioned causes include: increasing depth of exploitation, operation in fault zones, within the zones of protective pillars and near edges and debris.

The most active seismic was the area of the Kłodnica fault in the Main Saddle. Seismic activity in this region was mainly related to the mining operation carried out in ore and saddle layers. Rockbursts of the highest energy of 10^8 and 10^9 J occurred in the zone of the Kłodnica fault and the faults associated with the operation of the seams 418, 502, 507 and 510. In 2015, there was a rockburst of $4 \cdot 10^9$ J, which is the strongest phenomenon in the GZW.

Table 1. Output, number and consequences of rockbursts in GZW in 2007–2016.

Year	Total output, mln tonnes	Output form seams with tremor hazard, mln tonnes	Total energy of seismic tremors, J	**JWE, J/t**	Number of rock burst	**Rockburst structural unit**	Number of structural units in which rockbursts occurred	Total output in the mines where the rockburst occurred, mln tonnes
2007	87.40	44.60	2.31E+09	**26.41**	3	**Jejkowice and Chwałowice Trough (NJ) Main Saddle (SG)**	2	5.84
2008	83.60	41.90	2.01E+09	**24.04**	5	**Main Saddle (SG)**	1	14.25
2009	77.50	34.30	2.36E+09	**30.50**	1	**Main Saddle (SG)**	1	2.49
2010	76.10	35.80	4.48E+09	**58.84**	2	**Jejkowice and Chwałowice Trough (NJ)**	1	5.06
2011	75.50	34.20	1.95E+09	**25.80**	4	**Jejkowice and Chwałowice Trough (NJ) Main Saddle (SG) Bytom Trough (NB)**	3	10.80
2012	79.20	37.60	1.88E+09	**23.74**	1	**Jejkowice and Chwałowice Trough (NJ)**	1	2.75
2013	76.47	36.90	2.13E+09	**27.86**	1	**Bytom Trough (NB)**	1	1.04
2014	72.50	36.00	5.88E+09	**81.10**	1	**Jejkowice and Chwałowice Trough (NJ)**	1	4.88
2015	72.19	37.10	9.45E+09	**130.90**	2	**Main Saddle (SG)**	2	4.34
2016	70.40	36.90	2.90E+09	**41.19**	1	**Main Saddle (SG)**	1	0.74

Another area with high seismic activity was Bytom Trough, where seismic activity was associated with the operation of saddle and Poręba layers. In this area, very high seismic activity is observed. Outbreaks of events of energy greater than 10^8 J were located in the axis of the trough at a depth of several hundred meters below the mining activity.

In the Main Trough seismic activity, seismic activity occurred during the exploitation of the layered layers in which the most active seismic seams were 207 and 209. The strongest event of $3 \cdot 10^9$ J energy occurred in this area during operation in seam 207.

The large number of events recorded in Jejkowice Trough was caused by the operation of Jaklowice layers, mainly seams 703, 712 and 713.

Due to the inactivation of the mining works in Kazimierz Trough, low seismic activity was observed with a small number of rockbursts, representing only 0.1% of the other GZW structural units.

Despite the comparable values of total energy of tremors recorded within the area of Bytom Trough ($5.48 \cdot 10^9$ J) and Jejkowice and Chwałowice Troughs ($5.55 \cdot 10^9$J), the latter region is characterised by increased possibility of tremor due to the increasing number of tremors with higher energy values.

The energy analysis of tremors which caused rock bursts between 2001 and 2015, in relation to the number of tremors, shows that:

- Bytom Trough (NB) - 4 rock bursts of total energy Et = $4.02 \cdot 10^7$ J; average energy of a tremor that caused a rock burst amounted to about $1.01 \cdot 10^7$ J; 1,492 - total number of tremors
- Jejkowice and Chwałowice Trough (NB) - 10 rock bursts of total energy Et = $3.77 \cdot 10^8$J; average energy of a tremor that caused a rock burst amounted to about $3.77 \cdot 10^7$ J; 2,991 - total number of tremors.
- Main Saddle (SG) - 28 rock bursts of total energy Et = $4.78 \cdot 10^9$J; average energy of a tremor that caused a rock burst amounted to about $1.71 \cdot 10^8$ J; 8164 - total number of tremors.

Thus, in terms of rock burst hazard in the earlier mentioned regions, NB and NJ, the average energy of tremors that had resulted in a rock burst was comparable.

Probability of Rock Burst indicators: Bytom Trough (PT_{NB} =0.00268) and Jejkowice Trough (PT_{NJ} = 0.00334) are comparable despite the fact that the number of rock bursts and tremors in Jejkowice and Chwałowice Troughs was twice as high as in Bytom Trough.

Both seismic hazards such as rockbursts, are more accurately determined by the Regional Rock Burst Indicator (RWT), which is defined as the dependence of the aggregate energy of the tides versus the total energy of the seismic shocks of a given year.

Table analysis shows that Regional Rock Burst Indicator applies only to the three structural units of Upper Silesian Coal Basin. The maximum value is reached with respect to the recorded seismic activity and the number of tremors that had caused rock bursts in the area of the exploitation of mines located within Main Saddle (SG). The next regions include: Bytom Trough, Jejkowice Trough, and Jastrzębie Fold Trough, which have a significantly lower RWT value.

Summing up the above, it should be stressed that the research showed that between 2001 and 2015 rock burst hazard in GZW mines included only the three areas.

Estimated Regional Rock Burst Indicator, which indicates the level of rockbursts hazard, showed a relatively high value for the Main Saddle (RWT_{SG} = 0.234), which clearly indicates a very high risk of rockbursts in the area of this structural unit.

Bytom Trough (NB), and Jejkowice and Chwałowice Trough (NJ) are characterised by lower RWT values ($RWT_{NB,NJ}$ = 0.007).

4 Conclusion

The analysis of the tremors and rock bursts catalogue stored in GIG's (Central Mining Institute in Katowice) archive allowed to conclude that:

- despite the declining hard coal production in the last 10 years, the energy expenditure (JWE J/tonnes) has shown an upward trend by 2015,
- despite the decrease in the number of rockbursts, the total value of their seismic energy, especially high energy rockbursts > 10^5 J, remains at the JWE average level (47.04 J/tonnes) of the last ten years,
- in general, the activity was subject to change and depended on the intensity of mining works and diverse mining and geological conditions of the region,
- seismic activity occurred in five regions of GZW, representing the following structural units: Bytom Trough (NB), Main Saddle (SG), Main Trough (NG), Kazimierz Trough (NK), Jejkowice and Chwałowice Trough (NJ),
- only three are classified as areas susceptible to rock bursts, they include: Main Saddle (SG), Bytom Trough (NB), Jejkowice and Chwałowice Trough (NJ),

- the area characterised by the greatest seismic activity was Kłodnica Fault and associated faults in Main Saddle (SG). The maximum number of tremors was recorded in that region, including the events that caused rock bursts.
- another seismically active region characterised by a number of rock bursts is Jejkowice and Chwałowice Troughs (NJ), the smallest number of events was recorder in Bytom Trough (NB).
- in Main Trough (NG), despite high seismic activity, no tremors resulting in rock bursts were recorded.

Detailed studies on the impact of faults and the influence of fractures have shown their association with the risk of rock bursts and thus the seismic activity determined by the Unitary Energy Expenditure (J/t). These determinations were made based on the rockbursts statistics of the GZW coal mines.

References

1. B.H.G. Brady, E.T. Brown, *Rock mechanics for underground mining,* (University Press, Cambridge 1993)
2. E. Hoek, P.K. Kaiser, W.F. Bawden, *Support of underground excavations in hard rock* (Funding by Mining Research Directorate and Universities Research Incentive Fund, 1995)
3. Z. Pilecki, *Dynamic analysis of mining tremor impact on excavation,* Proc. Int. FLAC Symp. on Numerical Modeling in Geomechanics, Detournay & Hart, 397-400 (Minneapollis, USA 1-3 Sept. 1999)
4. J. Dubiński, W. Konopko, *Tąpania – ocena, prognoza, zwalczanie* (Wydawnictwo GIG, Katowice 2000)
5. J. Dubiński, Z. Pilecki, W. Zuberek, eds., *Badania geofizyczne w kopalniach – przeszłość, teraźniejszość i zamierzenia na przyszłość* (IGSMiE PAN, Kraków 2001)
6. Z. Burtan, Z. Rak, J. Stasica, *Priorities in the development of mining systems in the mining sector in Poland,* Gospod. Surowcami Min. **24**, 1, 185-200 (2008)
7. W. Konopko, R. Patyńska, *Warunki występowania tąpnięć w kopalniach węgla kamiennego*, Przegląd Górniczy **64**, 1, 12-17 (2008)
8. J. Brodny, *Determining the working characteristic of a friction joint in a yielding support*, Arch. Min. Sci. **55**, 4, 733-746 (2010)
9. J. Brodny, *Tests of friction joints in mining yielding supports under dynamic load*, Arch. Min. Sci. **56**, 1, 209-227 (2011)
10. J. Brodny, Arch. Min. Sci. **57**, 4 (2012) doi: 10.2478/v10267-012-0015-4
11. Z. Burtan, A. Zorychta, J. Cieślik, D. Chlebowski D., Arch. Min. Sci. **59**, 3, 691 – 704 (2014) doi: 10.2478/amsc-2014-0048
12. A. Goszcz, *Wybrane problemy zagrożenia sejsmicznego i zagrożenia tąpaniami kopalniach podziemnych* (Biblioteka Szkoły Podziemnej Eksploatacji Złóż, Wydawnictwo Nauka Technika, Kraków 2004)
13. Z. Pilecki, *Metoda oceny zachowania się masywu skalnego wokół wyrobiska podziemnego*, Studia, Rozprawy, Monografie **59** (IGSMiE PAN, Kraków 1999)
14. R. Patyńska, *Wpływ kierunku wybierania pokładu na jednostkowy wydatek energetyczny (JWE, J/t),* XXVIII Zimowa Szkoła Mechaniki Górotworu i Geoinżynierii 437-450 (Oficyna Wydawnicza Politechniki Wrocławskiej, Wrocław 2005)
15. R. Patyńska, K. Stec, *Aktywność sejsmiczna Górnośląskiego Zagłębia Węglowego a zagrożenie tąpaniami w latach 2001-2015,* Zeszyty Naukowe – Bulletin of the Mineral

and Energy Economy Research Institute of the Polish Academy of Sciences **97**, 135-144 (2017)

16. R. Patyńska, *Zagrożenie tąpaniami. Raport roczny o stanie podstawowych zagrożeń naturalnych i technicznych w górnictwie węgla kamiennego*, 88-100 (Wyd. GIG, Katowice 2008-2017)
17. K. Stec, *Zagrożenie sejsmiczne. Raport roczny o stanie podstawowych zagrożeń naturalnych i technicznych w górnictwie węgla kamiennego*, 72-88 (Wyd. GIG, Katowice 2008-2017)
18. R. Patyńska, K. Stec, *Regional Rockburst Indicator for structural units of Upper Silesian Coal Basin,* Studia Geotechnica et Mechanica **39**, 2, 32-45 (2017)
19. E. Pilecka, *Statystyczna analiza związku między położeniem epicentrów wysokoenergetycznych wstrząsów a lineamentami na obszarze Górnośląskiego Zagłębia Węglowego*, Gospod. Surowcami Min. **23**, 4, 101-109 (2007)
20. E. Pilecka, Z. Pilecki, *Analysis of relation between induced seismic activity and satellite data*. Proc. 19th Symposium on Application of Geophysics to Engineering and Environmental Problems SAGEEP, EEGS, 346-355 (Seattle, USA, 2-6 April 2006)

E3S Web of Conferences **24**, 02001 (2017) DOI: 10.1051/e3sconf/20172402001
AG 2017 - 3rd International Conference on Applied Geophysics

Change in the physical parameters of rocks due to hydrofracturing

Henryk Marcak[1,*]

[1]Polish Academy of Sciences, Institute of Geophysics, 01-452 Warszawa, Poland

Abstract. Hydrofracturing is a procedure performed for the activation and intensification of fluid flow in shale and geothermal deposits. This operation involves the risk of inducing strong seismic shocks and the leakage of borehole liquids into drinking water reservoirs. The changes of the physical properties of the rocks deformed by hydro-fracturing are modelled numerically. The structure of the model is presented in the paper. It is also demonstrated that the development of fracture zones resulting from the hydro-fracturing process can be investigated by electromagnetic and seismic methods. The flow of liquids through linear elements (fractures and cracks) increases the risk of destructive action through a combination of tectonic and hydro-fracture stresses.

1 Introduction

In the hydrofracturing process, high pressure liquid is introduced into the horizontal part of a borehole forcing the liquid against the rock and causing rock fracturing, cracking the rock surrounding it, until it finally facilitates the flow of reservoir fluids, in particular gas, into the borehole. According to IMS Global Insight and Drilling Info Inc., nearly half of the holes currently drilled in the US for oil and 2/3 for gas are hydro-fractured.

This method of hydrocarbon exploitation needs to be controlled. There are cases , rather rare, in which hydrofracturing produces strong seismic events. Examples include such seismic events as those which occurred during the hydro-fracturing of geothermal deposits in Basel [8]. Despite the fact that the injection of water into the rock had ceased and the water was even being pumped out, a number of seismic shocks, with increasing energy, were induced and threatened the city. During the hydrofracturing of wells in the Horn River Basin, Canada, a tremor of magnitude 3.8 was generated. There is also a threat to the public due to leakage of reservoir fluids into drinking water reservoirs as a result of hydrofracturing.

In this situation, it would be useful to have geophysical methods which allow one to control and predict the development of danger situations during hydrofracturing processes. However, geophysical methods allow one to recognise changes in the physical properties of rocks and a feasibility study on geophysical methods should be preceded by an analysis of the physical changes in the rocks resulting from hydrofracturing.

* Corresponding author: marcak@agh.edu.pl

2 Geomechanical aspects of the hydrofracturing process

Hydrofracturing is a result of introducing pressure exceeding the formation pore pressure into a horizontal part of a borehole for a sufficiently long period of time (Fig. 1).This effect is achieved by the installation of inflatable packers in the wellbore sections which close and seal them. Water under high pressure with sand and chemicals which is pumped through the packer for a long period of time produces a vertical fracture.

Two phases of the geo-mechanical deformations effecting changes in the physical properties of the medium are important. One of these is the formation of vertical fractures in the rock. The second is the spread of cracked zones around the vertical cracks. Both of these are discussed separately.

The Griffith model can be used in a simplified description of the first phase. Due to the symmetry introduced by a horizontal borehole, a two-dimensional description can be sufficient to describe the stresses in rocks under hydro-fracturing. The stress tensor in this system has three components σ_x, σ_y and $\tau_{x,y}$. By rotation of the coordinate system two main stresses can be estimated σ_I and σ_{II}.

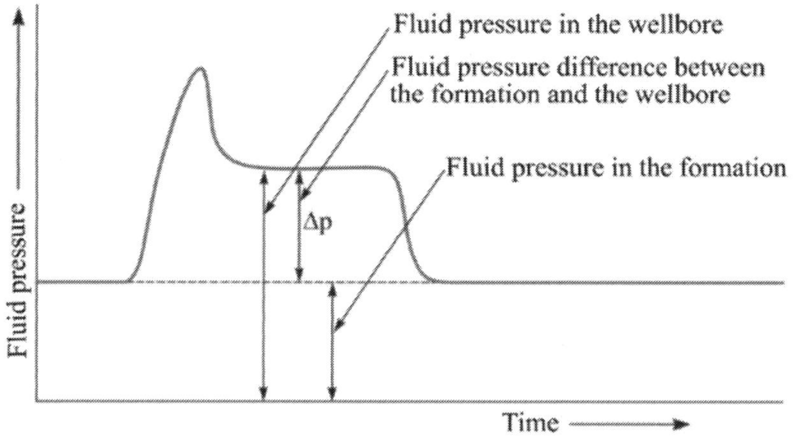

Fig. 1. Changes of pressure in rocks under hydrofracturing.

Mohr's circles are constructed in a coordinate system, ith axes for compressive stress σ and shear stress τ. Circles are drawn with centres on the σ axes at a point $(\sigma_I + \sigma_{II})/2$ and with radius $(\sigma_I - \sigma_{II})/2$.

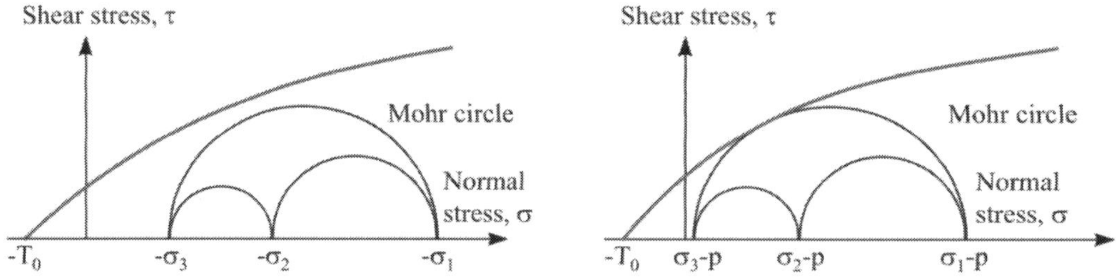

Fig. 2. Effect of hydraulic pressure on the change of stress in Mohr's circles and the Coulomb failure envelope.

The circle is shown in Figure 2a. The failure envelope (red line in Fig. 2) allows one to determine the shear stress which represents the limits of rock strength. If the medium is subjected to hydraulic pressure, the principal stresses in the graph are reduced by this value p. As shown in Figure 2b, this results in crack formation for the same values of principal stress that were too weak to produce cracking without hydraulic pressure.

Figure 3 shows the distributions of the radial, tangential and shear stresses around a circular opening in a two-axial stress system. The following values are assumed: vertical stress σ_z equal to 1000 MPa, Poisson's ratio ν equal to 0.25. In Fig. 3, an intensive increase of the tensile stress can be noted in the roof and bottom of the opening. This example explains why the crack extends in a direction perpendicular to the minimum main stresses.

There are many papers outlining the geo-mechanical conditions for extending fractures as a result of hydrofracturing. The classical one is the so-called two-dimensional model of Perkins-Kern-Norgren [13], [6]. A constant crack height is assumed in this model. The mathematical description of the dynamic development of the fracture permits its size, the pressure inside the fracture, and stresses at the tips of the fracture to be predicted as a function of time. In particular, the pressure at which breakage begins can be determined from the relationship:

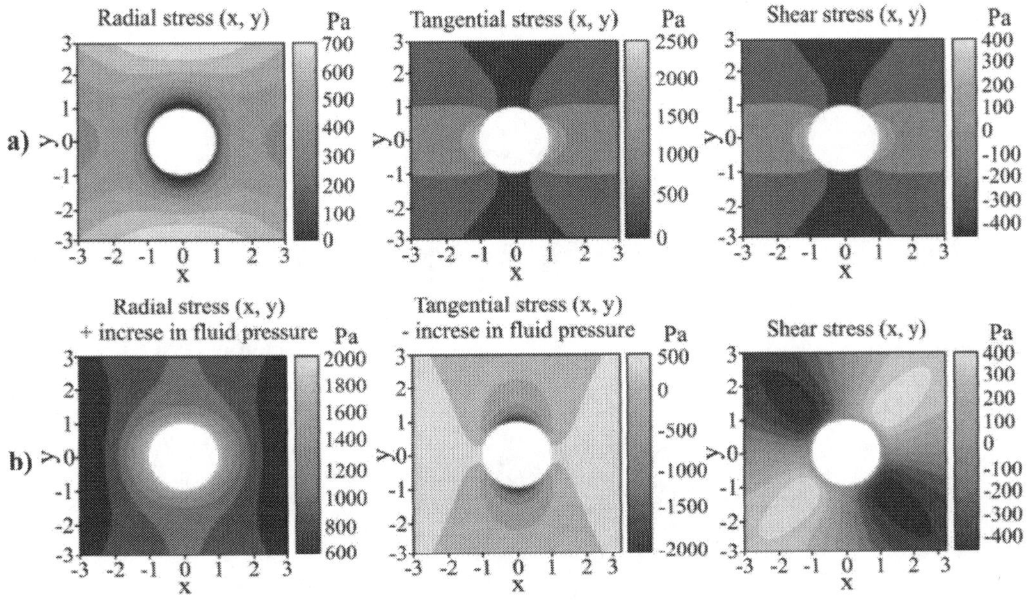

Fig. 3. Stress distribution around the circular opening before and after applying hydraulic pressure a) for a medium without porosity b) in a porous medium.

a) for a medium without porosity

$$p_b = 3\sigma_{min} - \sigma_{max} + T - p_V \qquad (1)$$

b) in a porous medium

$$p_b = \frac{3\sigma_{min} - \sigma_{max} + T - \alpha\frac{1-2\nu}{1-\nu}p_v}{2-\alpha\frac{1-2\nu}{1-\nu}} \qquad (2)$$

T – tensile strength of the rock, σ_{min}, σ_{max} – minimum and maximum main stress, p_v – porous pressure, α – Biott coefficient, ν – Poisson's ratio.

Half the length of the crack L and its width w can be estimated from the formulae:

$$L = 0.48\left[\frac{8GQ^3}{(1-\nu)\mu}\right]^{1/6} t^{3/2} \qquad (3) \qquad\qquad w = 1.32\left[\frac{8(1-\nu)Q^3\mu}{G}\right]^{1/6} t^{1/3} \qquad (4)$$

Q – injection ratio, G – stiffness coefficient, μ – hydraulic viscosity, t – injection time.

This mathematical model, supplemented by the estimation of energy losses resulting from the destruction of the medium around the developing fracture [18], allows one to estimate the fracture growth over time, as is shown in Figure 4 (average crack width is 14 mm).

Now we are coming to the second physically important, geo-mechanical phase of hydrofracturing. Both shear and tensile stresses in the zones surrounding the tips of an advancing fracture [4] can form a cracked zone.The following criterion of micro-crack appearance [15] can be used:

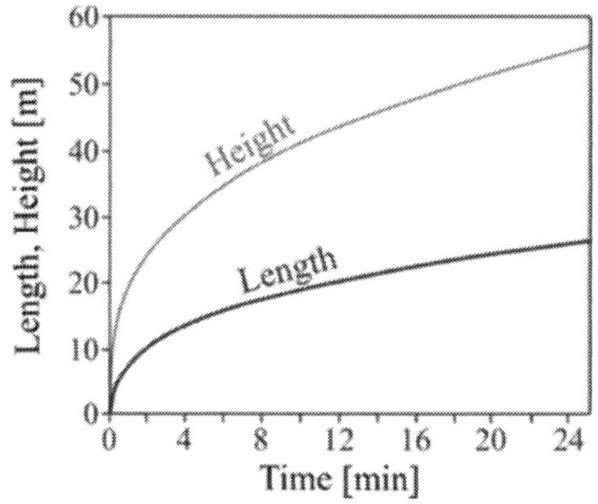

Fig. 4. Height and length of fracture as a function of time during hydro-fracturing [18].

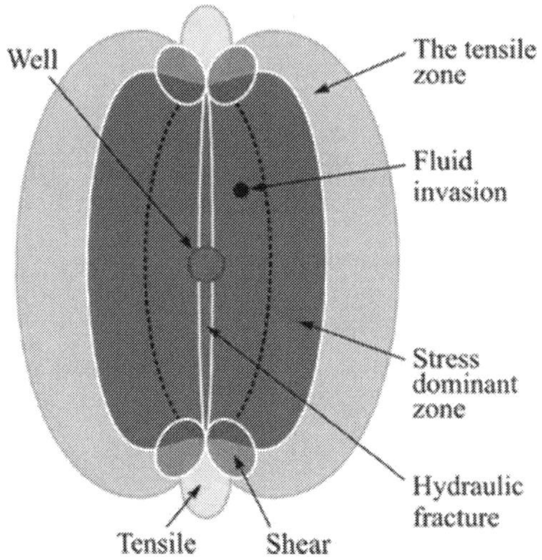

Fig. 5. Development of deformation around a vertical fracture.

$$\sqrt{\beta^{-1}\left(\sigma_t^2+\sigma_s^2\right)+\sigma_n^2} \geq T \qquad (5)$$

σ_s , σ_t and σ_n are stresses respectively shear , tangential and normal to the fracture surface, T is the tensile strength of the rock, β is a coefficient depending on the fracture length.

If a cracked zone is created around the main fracture then a rapid increase in hydraulic permeability can be expected in this zone. With some delay, this zone is saturated with fluid and the saturation changes the stress regime again.

The complicated distribution of stresses, as shown in Fig. 5, alters the physical properties. The differentiation of physical properties has consequences in the differentiation of the electrical resistance and seismic velocity of the rock. These can be recognised by geophysical methods. The distribution of the electromagnetic field was modelled in the surroundings of the hydro-frack and the results are presented in paper [9]. In modelling it was assumed that the porosity Φ linearly increases with pore pressure p, temperature T and stresses σ according to the following formula (i is the index of the grid element).

$$\partial\Phi_1 = (\frac{\alpha_1^2}{K_1} + \frac{\alpha_1 - \Phi_1}{K_s})\partial p_1 + 3\alpha_{T,1}\alpha_1\partial T_i + \frac{\alpha_1}{K_1}\partial\sigma_v \qquad (6)$$

α is Biota's coefficient, α_T is the thermal conductivity coefficient, and K_1 and K_s are the stiffness coefficients of the rock and rock skeleton.

The results of the modelling [9] allow one to assess the development of the electromagnetic field during fracture formation in the hydrofracturing process. If the pores are filled with brine, the results of the electromagnetic measurements are particularly informative.

Numerical modelling of the stresses in a pore-elastic medium has been proposed in [12]. It is based on solving differential equations by approximating them with difference equations and by using the specific capacitive parameters S_c and S_ε (formulae 7). The deformation of the rock can be predicted by solution of the parabolic equations under accepted boundary and initial conditions. These deformations can be transformed into changes of physical properties, primarily seismic velocity [3] and electrical resistance.

$$S_c = \rho g\frac{\partial p}{\partial\sigma} \quad S_\varepsilon = \frac{\partial p}{\partial\varepsilon} \quad S_c = \rho g S_\varepsilon \quad \frac{\partial p}{\partial t} = \nabla\left[\frac{\phi}{S_e}\nabla p\right] + B\frac{\partial\sigma}{\partial t} \quad \frac{\partial p}{\partial t} = \nabla\left[\frac{\phi}{S_e}\nabla p\right] + \frac{\alpha}{S_\varepsilon}\frac{\partial\varepsilon}{\partial t} \qquad (7)$$

Φ – permeability, B – Skempton coefficient.

The assumptions under which the models discussed above are constructed make them symmetrical in relation to the main fracture and fully controlled by the volume of injected liquid. To investigate how this model corresponds to reality, seismicity is analysed in the Geysers geothermal field in USA. The Geysers geothermal field is a very extensive deposit located in the subduction zone. A granite intrusion heats up the reservoir water to a temperature of 760 degrees. To stimulate its exploitation, water is injected into 60 boreholes and steam is exploited from 330 wells. A modern seismic network is operating in the area of the deposit, which records about 4,000 shocks per year with a magnitude between 1 and 4.5.

The results of seismic measurements are processed, mini-earthquake epicentres are located and their seismic moments are estimated. Such data allowed us to examine the distribution of the mini-earthquake epicentres in the area of the Parti 9 injection well. The seismicity recorded between 19 November 2007 and 31 August 2008 (total period) was related to the ratio of water injection. As shown in Figure 6d, from November 2008 to the beginning of April 2008 there was period with a stable injection ratio to the Parti 9 well. After this period, the volume of injected water continuously decreased (decrease period) until the end of the observation period. The relationship between the seismic activity (number of seismic events per unit time) and the volume of injected water ratio was established and can be approximated by the formula shown in Fig. 6c. The seismic activity tends to zero when the injection stops (injection is controlling seismicity). The distribution of mini-earthquake epicentres was analysed in two periods, "total" and "decreasing" (Fig. 6a and Fig. 6b).Two conclusions can be drawn from the distributions. The water channels are linear. Decreases in water injection result in a significant decrease in the number of mini-earthquake epicentres in the area, becoming significantly less intensive along the linear channels at larger distances from the well.

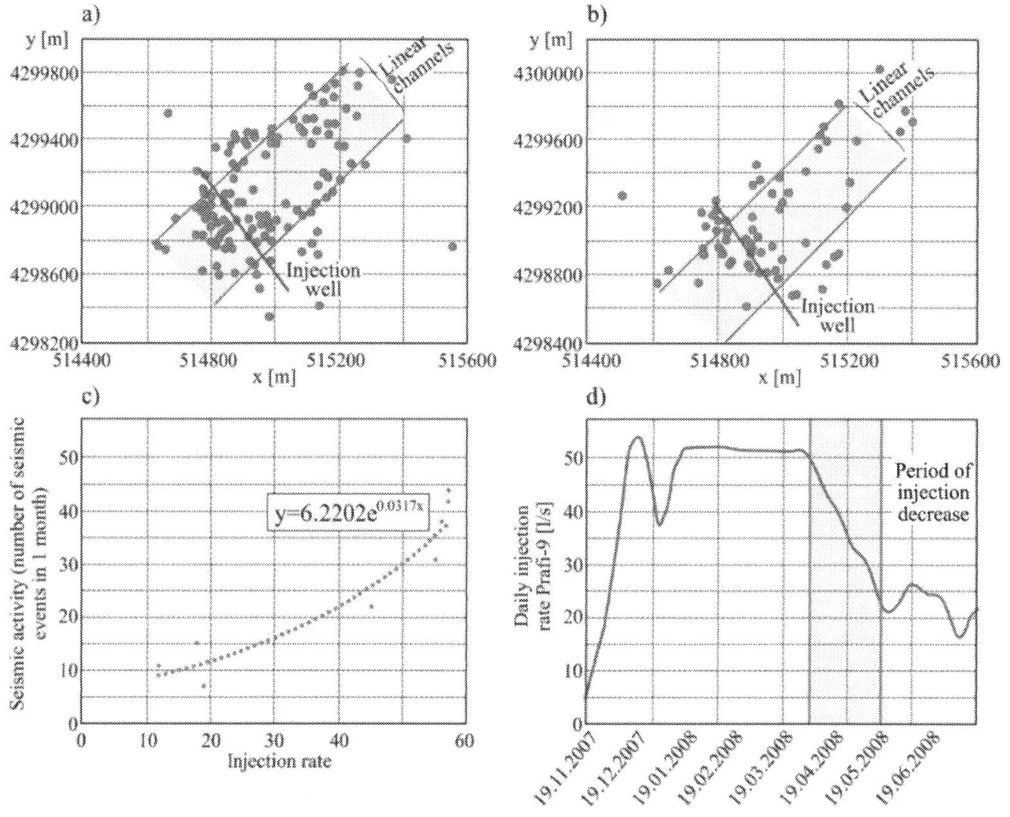

Fig. 6. Location of micro-earthquake epicentres induced by water injection into the Geysers field, USA: a) distribution of micro-earthquake epicentres registered in the period 19th November 2007-31st August 2008, b) distribution of micro-earthquake epicentres registered in a period of decreasing volume of water injection (marked in figure d), c) relationship between seismic activity and the volume of water injection ratio, d) change of water injection ratio with time.

The liquid flow along linear elements is different from that in volume. Hydraulic resistance to linear flow and its development over time can be described by the following formula from Bear et al. [2]:

$$\phi_i = \frac{1}{12}\frac{\rho g L}{\mu}\int_0^{\infty} b^3 f(b)db \qquad -\frac{\partial}{\partial x_i}\left(\phi_i \frac{\partial \varphi}{\partial x_i}\right) = S_i \frac{\partial \varphi}{\partial t} \qquad \varphi = z + \frac{p}{g\rho} \tag{8}$$

L – length of the linear element, b – width of the linear element described by probability distribution f (b), μ– viscosity of the fluid, Si – hydraulic capacity of the medium, z – depth, g – gravity constant, ρ– density, p – hydraulic pressure.

These equations indicate that, at a small hydraulic capacity of the geological medium, the hydraulic pressure at the ends of the linear elements, even at long distances, increases significantly over time, especially when the channels are widened by non-elastic deformations caused by hydraulic pressure.

It can be concluded that one needs more information to model stresses, deformations and fracturing as a result of hydro-fracturing in an anisotropic medium with linear hydraulic channels and so therefore it is practically impossible. Geophysical observations using seismology and seismic interferometery methods can be helpful in assessing the risk of the dangerous hydrofracturing developments.

Griffith's generalised theory may be helpful in interpreting the seismological observations. The total energy ΔF released by the formation of N cracking can be estimated as [18]:

$$\Delta F = Nw(-B^2\sigma^2\langle c^2\rangle + 4\gamma\langle c\rangle) \tag{9}$$

$\langle c\rangle$ is the expected value of a half–length of the crack, σ – the tensile stress in the element of medium B is constant depending on the elastic properties of the medium, w is the width of the block in which the cracks are formed and 2 γ is the energy needed to separate the cracked surface.

A coefficient $G = \frac{\partial \Delta F}{\partial A}$ of energy relaxation per unit increment of free fracture surface can be introduced, where A = 4cw. This parameter determines the rate of growth of the cracked area [10] and if it is near the critical value then its increase may be noticeable in the measurement data, both seismological and seismic [5]. Ouchi in [11], shows that an increase in seismic events dn (t) depends on the size of the population n(t) and the coefficients of growth and also depends on a h(t) representing local conditions:

$$dn(t)/dt = n(t)\left[\alpha - \beta n(t) - \int_{-\infty}^{t} n(s)h(t-s)ds\right] \tag{10}$$

$$R(t) = \frac{\int_{t_i-t_s}^{t_i+t_s} f_{obs}(t'+t_i)f_{ref}(t')dt'}{\int_{t_i-t_s}^{t_i+t_s} f^2{}_{obs}(t')dt' \int_{t_i-t_s}^{t_i+t_s} f_{ref}{}^2{}_{ref}(t')dt'} \tag{11}$$

Approaching a critical situation means an increase in the growth coefficient so that, despite the reduction of the main cause of the seismicity, the volume of injection fluid, seismic activity does not decrease or even increase. On the other hand, with the seismic interferometry method, the seismic velocity distribution in the geological medium can be recognised from the seismic noise recorded.

If the medium is under significant non-elastic deformation over time, then changes in the seismic velocity distribution are directly related to the deformations. The stretching coefficient (11), [7] the maximum of the correlation function calculated from the seismic noise recorded in different time windows, can control the deformation caused by hydro-fracturing.

3 Summary

The deformation process in a geological medium due to hydrofracturing is complicated. However some of its phases can be modelled on the basis of the algorithms presented in the article. Such modelling can be useful in planning and arranging the geophysical measurements, such as those obtained by seismological and seismic interferometry methods. The results of such measurements, together with data from other methods, can help in controlling the risk of dangerous rock deformations due to hydrofracturing.

This work was carried out in the framework of the Shale Gas Exploration and Exploitation Induced Risks (SHEER) project funded by the European Union's Horizon 2020 Research and Innovation Programme under grant agreement no. 640896.

References

1. BCOGC, *Investigation of induced seismicity in the Horn River Basin* (British Columbia Oil and Gas Commission open report, 2012)
2. J. Bear, C.F. Tsang, G. De Marsily, *Flow and contaminat transport in fractured rock* (Academic Press, 1993)
3. C. Brisco, M. van der Baan. *A review of seismic velocity response to variatiation in pore pressure pore-saturating fluid and confirming stress* (GeoConvection Calgary, Canada, 2016)
4. C. Cipolla, S. Maxwell, M. Mack, R. Downie, *A practical guide to interpreting microseismic measurements* (SPE North American Unconventional Gas Conference and Exhibition, 2011)
5. R. Czarny, H. Marcak, N. Nakata, Z Pilecki Z. Isakow, Pure. Appl. Geophys. **173**, 6, 1907-1916 (2016) doi 10.1007/s00024-015-1234-3
6. J. Geertsma, J.F. de-Klerk, J. Petrol. Technol. **21**, 12, 1571-1581 (1969) doi 10.2118/2458-PA
7. C. Hadzioannou, E. Larose O. Coutant P. Roux, J. Acoust. Soc. Am. **125**, 6, 3688–3695 (2009) doi: 10.1121/1.3125345
8. G. Hillers, S. Husen, A. Obermann, T. Planes, E. Laroce, M Campillo, *Noise-based monitoring and imaging of a seismic transient deformation induced by the 2006 Basel reservoir stimulation*, Geophysics **80**, 4, 51-68 (2015)
9. J. Kim, E. Schankee Um, J.G. Moridis, *Fracture propagation, fluid flow, and geomechanics of water-based hydraulic fracturing in shale gas systems and electromagnetic geophysical monitoring of fluid migration* (SPE 168578, Hydraulic Fracturing Technology Conference, Texas/USA, 2014)
10. I. G. Main, *A modified Griffith criterion for the evolution of damage with a fractal distribution of crack lengths: application to seismic event rates and b-values*, Geophys. J. Int. **107**, 353-362 (1991)
11. T. Ouchi, Pure. Appl. Geophys. **140**, 1, 15-28 (1993) doi 10.1007/BF00876868
12. V. Palciauskas, P.A. Domenico, Water. Resour. Res. **25**, 2, 203-213 (1989) doi 10.1029/WR025i002p00203
13. T.K. Perkins, L.R. Kern, J. Petrol. Technol. **13**, 9, 937-949 (1961) doi 10.2118/89-PA
14. J. Perrin, T. Cook, *Hydraulically fractured wells provide two-third of US national gas production US Energy* (Independent Statistics and Analysis U.S. Energy Information Administration, 2016)
15. G. Ruiz, M. Ortiz, A. Pandolfi, *Three-dimensional finite-element simulation of the dynamic Brazilian tests on concrete cylinders*, Int. J. Numer. Meth. Eng. **48**, 963-994 (2000)
16. D.R. Schmitt, M.D. Zoback, Int. J. Rock. Mech. Min. **26**, 6, 499-506 (1989) doi 10.1016/0148-9062(89)91427-7

17. C.H. Scholz, *The mechanics of earthquakes and faulting* (Cambridge University Press, 1990)
18. P. Valko, M.J. Economides, *Propagation of hydraulically induced fractures – a continuum damage mechanic approach,* Int. J. Rock. Mech. Min. **31**, 3, 221-229 (1994)

E3S Web of Conferences **24**, 02002 (2017) DOI: 10.1051/e3sconf/20172402002
AG 2017 - 3rd International Conference on Applied Geophysics

Comparative analysis of rock porosity in a selected region in the Carpathians based on laboratory data and well logs

Maciej Barnaś[1]* and *Edyta* Puskarczyk[2]

[1]Mineral and Energy Economy Research Institute, Polish Academy of Sciences, 31-261 Krakow, Poland
[2]AGH University of Science and Technology, 30-059 Krakow, Poland

Abstract. A comparative analysis of porosity is presented in the article. This was calculated from laboratory measurements and well logs for rocks from a selected region of the Carpathians. Analysis was performed for samples of sandstones, mudstones, shales and carbonates from four boreholes drilled in the Carpathian Foredeep region near Krakow, Poland. In laboratory measurements, porosity was assessed using two methods: helium pycnometry and nuclear magnetic resonance (NMR). Porosity from well logging was calculated from sonic profiling (SPHI). Cross-plots were used, which allowed for the assessment of the degree of correlation between the laboratory and well log data. The results obtained were analysed paying special attention to the relationship with geological structure. Based on the lithological criterion, the correlations turned out to be strong. The repeatability of the results obtained in different wells was verified. The results of the measurements are significant in relation to estimating porosity from well logs using laboratory data.

1 Introduction

The aim of the work is to compare the porosity of rocks from a selected region in the Carpathians calculated from laboratory methods and from well log SPHI. An evaluation is made of the degree of correlation between the results of laboratory measurements and well logs, based on laboratory measurements of porosity and sonic well log SPHI. Closely related topics were considered by Hartmann et al. [1] and Mason et al. [2], and in Poland (including Carpathian rocks) by Jarzyna [3], Gąsior and Przelaskowska [4] and Krakowska [5].

Well logs are a trustworthy source of data about physical parameters and variability measurements. However they are usually performed at sampling intervals of 10-25 cm. The measured values, even after applying corrections for the borehole, drilling mud, adjacent layers etc., sometimes can diverge from the real values due to poor vertical resolution and small radius of investigation. Laboratory measurements on samples yield accurate results but have a point-based character, therefore can only act as reference results in addition to well

* Corresponding author: maciej.michal.barnas@gmail.com

logs. Nevertheless it is necessary to underline that laboratory measurements can be performed in conditions that simulate those of the actual deposit (appropriate temperature and pressure).

The characteristics of the survey area, materials and methods are discussed in the initial part of this article. Afterwards the results of the correlation, well logs and laboratory data are outlined.

2 Survey area characteristics

The results from four boreholes (denoted as A-1, A-2, A-3 and A-4), drilled in the Carpathian Foredeep in the south-east Poland, were used for the analyses. The boreholes were located at a distance of between 400 and 2000 metres from each other, and were drilled into the same deposit. The deepest well reached a depth of 3000 m, the remaining three were drilled to depths of 1900 – 2250 m. Miocene deposits (composed of mudstones, claystones and sandstones) and Late Jurassic deposits (limestones and dolomites) were present. Two boreholes were drilled Middle Jurassic deposits (sandstones and limestones) and Permian-Triassic strata (mudstones, claystones and sandstones with anhydrite content). Well No. A-4 was drilled into Carboniferous limestones [6].

3 Materials and methods

In the case of the SPHI log, described among others by Jarzyna et al. [7], rock porosity is calculated from the Wyllie (1) or Raymer-Hunt-Gardner (2) equations:

$$SPHI = \frac{\Delta t_{log} - \Delta t_{ma}}{\Delta t_f - \Delta t_{ma}} \tag{1}$$

$$SPHI = \frac{\Delta t_{log} - \Delta t_{ma}}{\Delta t_{log}} \tag{2}$$

where:
Δt_{log} – travel time measured using sonic log,
Δt_{ma} – travel time in rock matrix,
Δt_f – travel time in fluid.

The porosity of rocks was assessed using two laboratory methods: helium pycnometry and nuclear magnetic resonance (NMR). During the pycnometry measurements, a sample was placed in a chamber with a calibrated volume and a gas (usually helium) is injected in. The difference between the volumes of the chamber and the gas provides the volume of the solid rock component. With the known mass and volume of the sample, it is possible to calculate the density. In the case of bulk density measurements, substances which cannot fill the smallest diameter pores are used. For the samples analysed in the article, equipment produced by the American Micromeritics company was used and that diameter was set at 25 nanometers. With density and bulk density, it is easy to calculate effective porosity [8].

Nuclear magnetic resonance uses the phenomenon of the effect of the magnetic field on hydrogen atoms [9]. The results of the surveys are expressed as relaxation time, which is affected by pore size, type of porous medium, movement, water content sealed by clay and that enclosed in capillaries, permeability, residual hydrocarbon saturation and the viscosity of crude oil [8].

Using the available data, descriptive statistics analyses were prepared based on a) wells and b) lithostratigraphical affiliation. Correlation dependences were also checked, and cross-plots were made. A sheet of correlation dependences was developed using our own Python scripts, using libraries NumPy and SciPy [10], pandas [12] and Matplotlib [13]. Scripts were performed based on source codes which were placed in the thesis [14] and were shared on the Internet platform GitHub [14].

The first script (used both in laboratory data analyses and by arranging them with the well logs) imports from .csv files, laboratory data – depth of collecting, values of porosity from two methods, borehole from which the sample originated and lithostratigraphical layer from which the sample was collected. Afterwards, using the least squares criterion, a linear regression along with the coefficient determinant R^2, was calculated. The drawing of cross-plots was the last stage of the process. The second script was used for importing well log values from .las files and deleting error values. As the results of laboratory measurements have a point character and well logs are continuous, the script arranges well log data with the corresponding laboratory data and saves output results to a .csv file, which can be used in the first script.

78 cylindrical samples, 10 – 15 cm in height, were available, including:

- 19 for well A-1, from a depth range of 1591 – 2245 m
- 6 for well A-2 well, from a depth range of 2000 – 2045 m
- 17 for well A-3 from a depth range of 1698 – 1816 m
- 36 for well A-4 from a depth range of 1678 – 2997 m.

Regarding the geological structure, the number of samples from each lithostratigraphical layer was as follows:

- Neogene – 9
- Late Jurassic – 46
- Middle-Jurassic – 6
- Permian-Triassic – 7
- Carboniferous – 12.

4 Results and analyses

4.1 Helium pycnometry and nuclear magnetic resonance

The total porosity of the core samples was measured twice using helium pycnometry and nuclear magnetic resonance. The correlation plot is shown in Fig. 1. The correlation is very good and the coefficient R^2 is very high (86% for the whole dataset). Three outlying samples were observed, with significant deviations from the regression line. They come from different boreholes (A-1 and A-2), and from different lithostratigraphic units (Neogene and Late Jurassic). Probably these two factors did not affect the results. For these samples, much higher total porosity was obtained from the NMR measurement. This is a confirmation of the trend observed for the entire set of samples – the value of the NMR porosity is greater than that of the helium pycnometry. It is most likely that this is due to the presence of bound water or isolated pores that are observable during NMR measurements. The best fit is for wells A-4 and A-1 (excluding one sample).

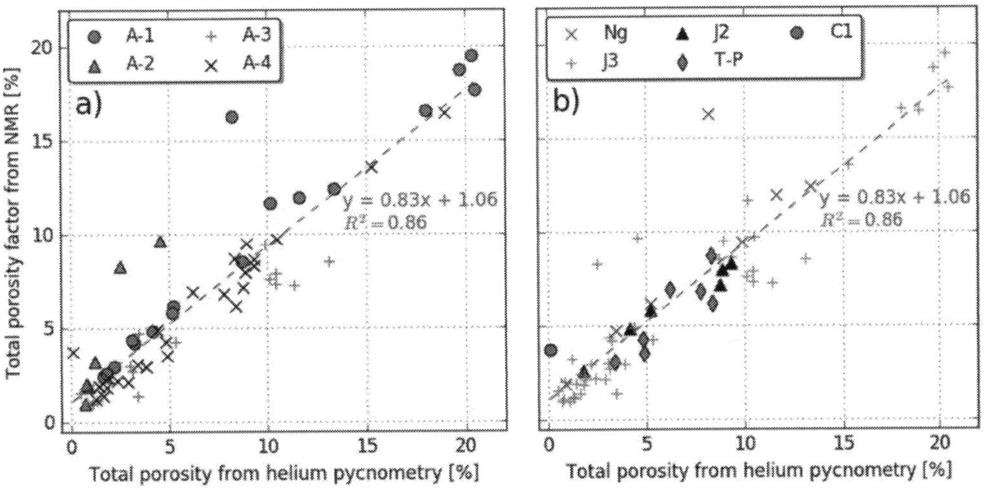

Fig. 1. Correlation of total porosity determined by pycnometer and using nuclear magnetic resonance. The shape and colour of the points differ due to the origin of the sample from a given well (a) or the lithostratigraphic unit to which it belongs: Ng – Neogene, J3 – Late Jurassic, J2 –Middle Jurassic, T-P – Triassic-Permian, C1 – Carboniferous (b); vertical scale common to both drawings.

4.2 Porosity parameter vs. porosity

The relationship of the porosity parameter P_p to porosity p is widely used [16]. Fig. 2 shows the log-log comparison. Using the formula (3), the pore structure index m was calculated:

$$P_p = A \cdot p^{-m} \tag{3}$$

where:
P_p – porosity parameter
p – porosity [%]

The size of m increases with the complexity of the pore structure and the degree of cementing.

The porosity parameter was calculated by measuring the electrical resistance of the cores. Its value is compared with the porosity parameter. For the pycnometry results, the calculated m index was 1.09, with a determination coefficient of 78%. However, for measurements using nuclear magnetic resonance, the calculated m value is 1.35, with a determination factor of 72%. In both cases the samples from the A-1 borehole have the best fit to the regression line. The presence of samples which diverge from the regression line can be the result of changes in pore diameter and pore size [16]. As far as lithostratigraphic division is concerned (Fig. 2b), only the Carboniferous has a small coefficient of determination. The other lithologies have a very good match (coefficient of determination between 76 and 95%) and the index of pore structure varies from 1.02 to 1.78.

DOI: 10.1051/e3sconf/20172402002

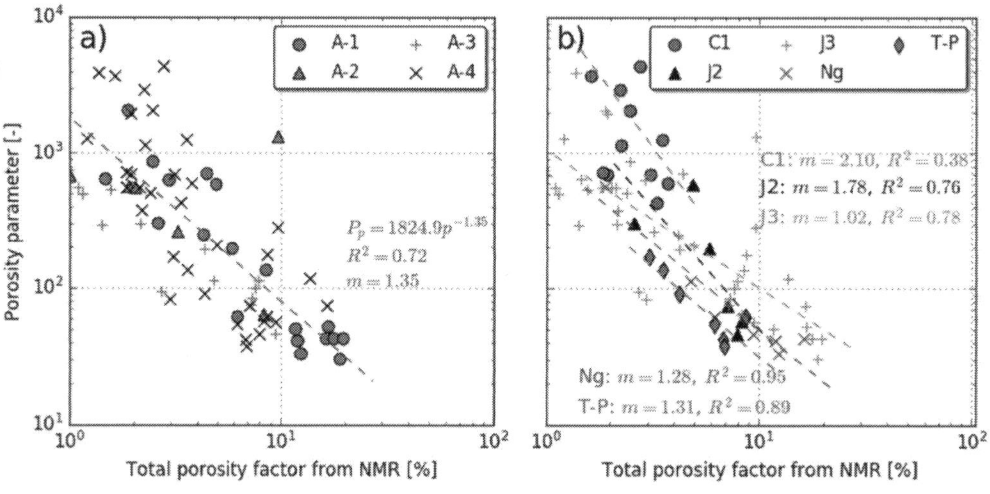

Fig. 2. Determination of the porosity parameter obtained from measurements using nuclear magnetic resonance. The shape and colour of the points are different due to the origin of the sample from the well (a) or the lithostratigraphic unit to which it belonged: Ng – Neogene, J3 – Late Jurassic, J2 – Middle Jurassic, T-P – Triassic-Permian, C1 – Carboniferous (b); vertical scale common to both drawings.

4.3 Comparison of porosity determined from laboratory measurements and the SPHI well log

The last stage was the comparison of laboratory tests with corresponding well logs. For this purpose, a SPHI log was used. It shows porosity calculated on the basis on either of the two equations: (1) or (2). In the case of the presence of oil or gas, adjustments need to be made in equations, as they increase the measured travel time.

Considering the above formulae, it is necessary to set the travel time in the rock matrix. In the case of laboratory measurements, the travel time is determined for a specific rock matrix. In the case of well logs, the travel time can be determined for a few rock matrix. This effect can cause discrepancies between SPHI and laboratory measurements, shown in the Fig. 3.

Fig. 3 shows the comparison of SPHI with porosity from laboratory measurements using pycnometry (a, b) and NMR (c, d). Only points from the Late Jurassic, with SPHI values greater than zero, were used to calculate the linear regression. The correlation is quite good – the coefficient of determination varies between 58 and 61%.

4.4 Comparison of data from well logs and laboratory measurements

Fig. **4** shows the comparison of the results of other logs with the results of laboratory measurements, after depth correction, for the specific interval at the border of the Neogene and the Jurassic. It is evident that the results of density measurements with the density log (RHOB) and porosity with the SPHI, porosity from density (DPHI) and neutron porosity (NPHI) logs are the best matched. For electrical logs and radioactive content, there are some differences between the laboratory data and well logs, while the relative variations of these values overlap.

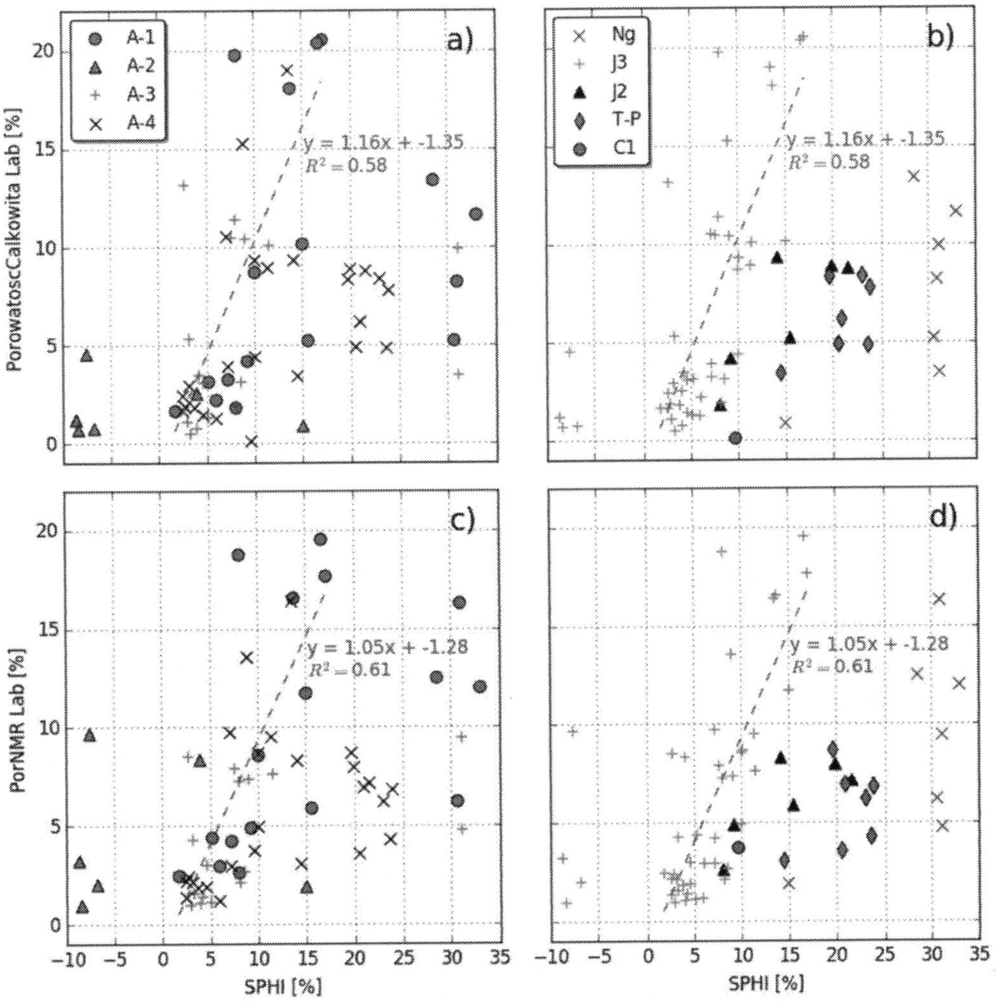

Fig. 3. Comparison of SPHI log results and porosity results measured using pycnometry (a, b) and NMR (c, d). Linear regression was calculated using only the results for Upper Jurassic samples for which the SPHI value is greater than zero. The shape and colour of the points differ due to the origin of the sample from a given borehole (a, c) or belonging to a lithostratigraphic unit: Ng – Neogene, J3 –Late Jurassic, J2 – Middle Jurassic, T-P – Triassic-Permian, C1 – Carboniferous (b,d).

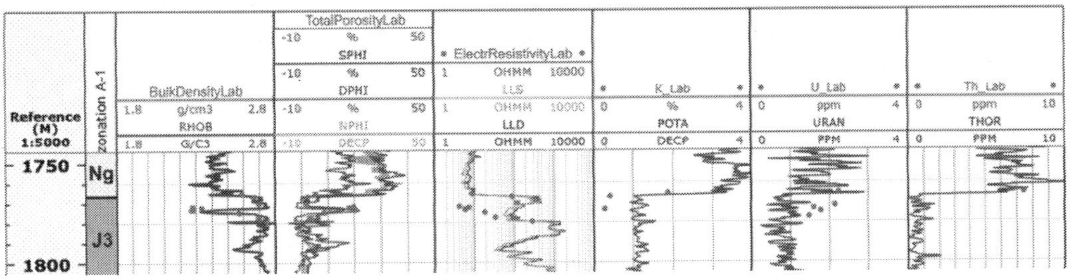

Fig. 4. Comparison of the results of well logs (curves) and laboratory data (red points) for selected intervals from well A-1.

5 Summary

The aim of the study was to carry out a comparative analysis of the results of laboratory measurements and well logs. In order to ensure the right amount of data for the analysis, four boreholes were chosen close to each other with similar lithostratigraphy. Cross-plots were made using Python scripts, along with libraries for statistical calculations. They were used to verify not only the existence and power of correlation between the two methods, but also to evaluate the effect of the origin of the sample from the lithostratigraphic unit and the reproducibility of the results in different wells.

In the case of total porosity, the results of two laboratory measurements (helium pycnometry and NMR experiment) were compared. A very strong correlation demonstrates the effectiveness and repeatability of both. The last step was the analysis of the laboratory data compared to the well logs – in this case SPHI. The correlation is good, but only for individual lithostratigraphic units. This is due to the calculation of the SPHI value, where it is necessary to set the interval time in the rock matrix. Accepting one value for the entire well has the effect that for some intervals the results will be close to the real porosity, and for others it will vary greatly from one to the other.

In conclusion, laboratory measurements are a very good addition to the well logs, and may be useful for both correlation and calibration of well logs. However, the vertical resolution of the methods being compared and their results and other possible discrepancies resulting from the measurement method need to be taken into account. When correlating the results and their interpretation, knowledge of the method of measurement and calculation of the quantities analysed is needed, and hence also of the potential differences in the results of the measurement.

The authors thank PGNiG SA for providing geological information, well logs and laboratory data. The analyses used in the article were undertaken as a part of Maciej Barnas' Master's of Sciences Thesis.

References

1. A. Hartmann, V. Rath, C. Clauser, Int. J. Rock Mech. Min. **7-8**, 42, 1042-1055 (2005) doi: 10.1016/j.ijrmms.2005.05.015
2. H. E. Mason, M. M. Smith, Y. Hao, S. A. Carroll, Enrgy Proced. **63**, 3089-3096, doi: 10.1016/j.egypro.2014.11.332 (2014)
3. J. Jarzyna, *Analysis of density and porosity measurements on cores and results of density log and Estymacja calculations in the selected wells in the West Carpathians*, Geologia **4/1**, 33, 39-58 (in Polish) (2011)
4. I. Gąsior, A. Przelaskowska, Acta Geophys. **4**, 62, 785-801 (2014) doi: 10.2478/s11600-014-0204-y
5. P. Krakowska, *Determination of reservoir potential of sedimentary rocks of the Precambrian and Palaeozoic age on the basis of the results of laboratory tests and well logs,* PhD Thesis (In Polish) (AGH University of Science and Technology, Kraków 2014)
6. A. Indyk, I. Molenda, P. Pasek, J. Florczak, D. Czarnecka, A. Naleźniak, *The results of interpretation well logs performer using DDL-D Halliburton set (final report), well: A-4 (name of the well was modified)* (unpublished, in Polish) (Geofizyka Kraków Sp. z o.o. 2011)
7. J. Jarzyna, M. Bała, T. Zorski, *Methods of well logging – acquisition and interpretation (in* Polish), (Wyd. AGH, Kraków 1999)

8. J. Zalewska, D. Cebulski, M. Dohnalik, A. Fuks-Kopiejka, J. Kaczmarczyk, S. Kowalska, A. Przelaskowska, J. Kiernicki, *Specialized service research for the well: A-4 (name of the well was modified), Task I: Specialized measurements of geophysical parameters of cores and muds with interpretation* (unpublished, in Polish) (Oil and Gas Institute – National Research Institute, Kraków 2011)
9. E. Puskarczyk, *Assessment of reservoir properties of rocks using the nuclear magnetic resonance experiment,* PhD Thesis (in Polish) (AGH University of Science and Technology, Kraków 2011)
10. P. Janusz, *Crossplots as an instrument of well logging interpretation. Bachelor's Thesis,* (in Polish) (AGH University of Science and Technology, Kraków 2016)
11. S. van der Walt, C. Colbert, G. Varoquaux, Comput. Sci. Eng. **13**, 22-30 (2011) doi:10.1109/MCSE.2011.37
12. W. McKinney, *Data Structures for Statistical Computing in Python*, Proc. of the 9th Python in Science Conference (2010)
13. J.D. Hunter, Comput. Sci. Eng. **9**, 3, 90-95 (2007) doi:10.1109/MCSE.2007.55
14. M. Barnaś, *Comprehensive analysis of laboratory data and well logs for selected reservoir,* Master's Thesis (in Polish) (AGH University of Science and Technology, Kraków 2016)
15. M. Barnaś, *ArticleWOC2017*, GitHub repository: https://github.com/barneygf/ArticleWOC2017 (access: 2017-09-21) (2017)
16. J. Zalewska, D. Cebulski, *Results of laboratory measurements of electrical properties of rocks,* Nafta-Gaz **4**, 66, 251-256 (in Polish) (2010)

Deformation of the rock pillar caused by bed operation using the room and pillar system

Wiesław Grzebyk[1], and *Lech* Stolecki[1*]

[1]KGHM Cuprum Ltd Research and Development Centre, 53-659 Wroclaw, Poland

Abstract. The copper ore deposit of the Legnica-Głogów Copper Mining District a one-deck bed formed from the contact of weak sandstone rocks, and solid, durable dolomite rocks, separated by a layer of copper slate. The bed operation is carried out using the room and pillar system. The course of deformation in such bed operation systems is determined by the mutual interaction of particular elements of the geo-mechanical system which include: the roof slab, the pillars and the excavation bed. Due to the possibility of human intervention, structural pillars – whose strength properties can be determined by the appropriate selection of their dimensions and shape – play a particular role in this system. The optimization of the properties of these pillars is mainly based on the results of laboratory tests of rock samples in the strength test machine and on the basis of visual observations of their behavior in situ. The article presents the experimental results carried out under pit conditions in order to determine the rock pillar reaction. Suitable observations were made using the so-called volumetric method. The obtained results were discussed in relation to the results of strength tests on rock samples, indicating the limited similarity to the deformation characteristics.

1 Introduction

One of the basic geomechanical problems in room and pillar systems of bed operation is the selection of the appropriate dimensions for structural pillars. At the Legnica-Głogów Copper Mining District's copper ore beds, the selection of these dimensions is conditioned by two factors: the guarantee that pillars will provide adequate support for the floor layers and leading the pillars to the subcritical state [1]. Due to technological considerations associated with the provision of the bed, many pillars in their original form have large dimensions, far exceeding the dimensions of the operating pillars at the front of the extraction works.

Generally, the actual operation of the structural pillars subjected to operating pressure is described by determining the deformation-to-strain characteristics obtained from laboratory measurements on rock samples [2-6]. The main research problem consists in demonstrating the consistency of the obtained results of the laboratory tests in this field with the actual deformation characteristics of pillars in mining conditions. The results of this study attempt to answer the above question.

* Corresponding author: lstolecki@cuprum.wroc.pl

2 Strength characteristics of rock samples in pre- and post-critical conditions

The equivalent of the structural pillar deformation process in the room and pillar operating system is usually the full deformation-to-strain characteristics of the rock sample obtained under uniaxial compression conditions. Thus, the obtained graph of linear strength for fragile rock samples always contains its ascending, peak and descending sections, which correspond respectively to the pre-destructive condition, the moment of destruction, and the post-destructive state of the rock. The mechanism of destruction for a fragile rock, in the context of linear and volumetric deformations, was provided by Bieniawski [2].

The results of laboratory tests that are carried out on rock samples with normalized shapes and dimensions and the loading rate (ISRM recommendation, Ulusey, Hudson, eds. 2007) mainly show a comparative advantage. On the other hand, the results of strength tests on samples of different dimensions and different load rates, i.e. parameters approximating the real conditions, are more useful for practical purposes. An exemplary effect of changing these parameters based on the mechanical characteristics is shown in Figures 1a and 1b. According to the results of these studies on sandstone samples, "the increase in deformation speed of the rock model results in an increase of its strength and rigidity in the pre-critical state and an increase of the rigidity module in the post-critical state, while the change of the geometry and dimensions of the model in the direction of its length increase relative to its height results in a decrease in the post-critical rigidity [7]."

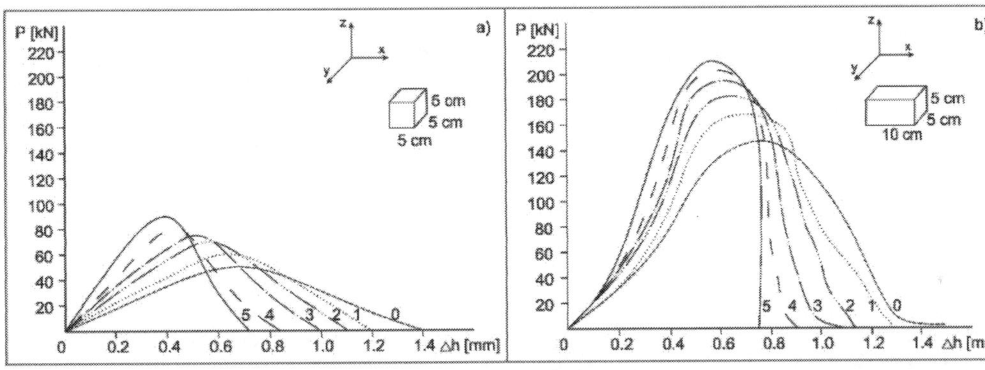

Fig. 1. Mechanical characteristics of square (a) and rectangular (b) models of clay sandstone for different load ing rates: 0 – 1, 1 – 10, 2 – 102, 3 – 103, 4 – 104, 5 – 105 μm/s (acc. to [7]).

3 Object-oriented research in situ

Recognition of the reaction of a pillar-type rock medium under the influence of increasing operating pressure was carried out in one of the operating areas of the copper ore beds of the Legnica-Głogów Copper Mining District. The bed lies at a depth of approx. 300 m to approx. 1300 m and is formed as a flat deck with a slight angle in the NW direction. The bed is in contact with solid and durable carbonate rocks (in the upper part) and much weaker sandstone rocks of Rotliegend, separated by a thin layer of copper slate. Operation of the bed is carried out in the one-stage room and pillar system with floor slab deflection.

Relevant tests were performed for the selected pillar positioned ahead of the operating front (Figs. 2, 3). The pillar has a triangular shape measuring 40 x 32 x 32 m and a height of approx. 4 m.

Fig. 2. Mining situation at the start of observation.

Fig. 3. Mining situation at the end of observation.

In order to determine the deformation characteristics of the pillar, the so-called volumetric measurement method [8] was used. To do this, special measuring probes were inserted into two horizontal holes in the sandstone and dolomite pillar sections; the basic elements of these probes are resilient cylindrical shaped vessels. After inserting the probes to a certain depth in the openings, pressurized hydraulic oil was introduced into the vessels, causing them to expand sideways and close contact against the wall of the hole. The construction of the probe prevented the tank from deforming along the axis of the hole. Location of research holes in the pillar in the horizontal projection is shown in Figs 2 and 3, whereas diagram of the measuring system in the vertical projection in Fig. 4. Pipes were routed from the probes, outwards from the boreholes, and connected to the absolute pressure sensors. Changes in oil pressure made at set time intervals were measured.

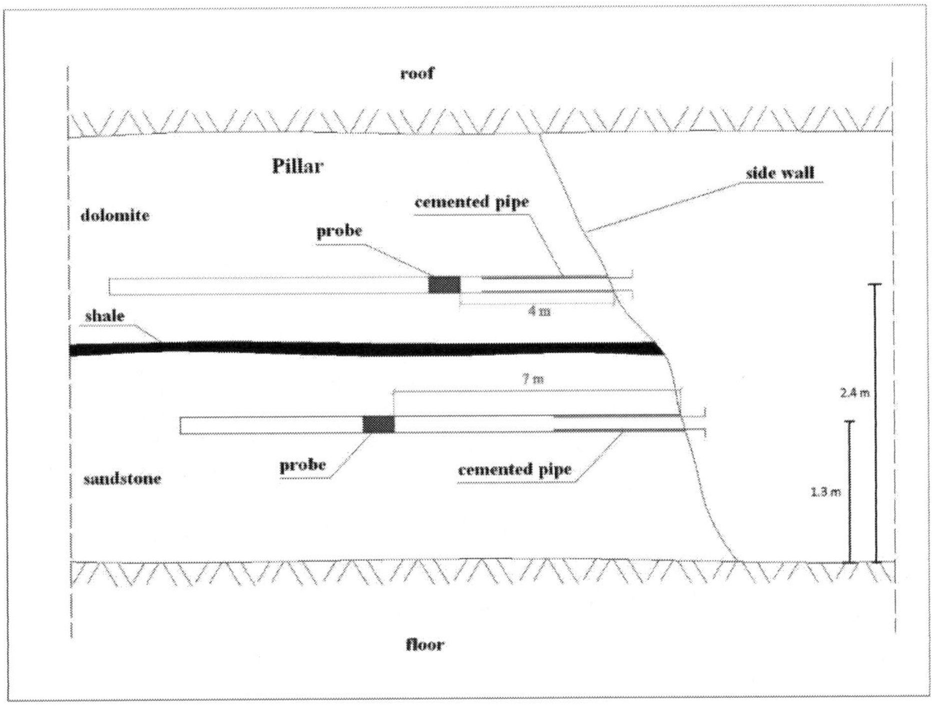

Fig. 4. Diagram of the measuring system in the vertical projection.

4 Measurement results

The described measurements were conducted over a period of 14 months. At the time of start of theirs, the operating front was approximately 240 m from the measuring station and over time, with relatively constant speed (approx. 6.31 µm/s) it approached the pillar. In the final stage of the measurements, the pillar was cut by creating a working in it, maintaining the efficiency of the measuring system. Measurements were completed when the operating front was approx. 12 m behind the pillar.

The obtained results of the observations in the form of graphs of pressure changes measured in the sandstone and dolomite parts of the pillar and the progress of the operating front are shown in Figure 5. The time axis, in such an assumed coordinate system, should be regarded as a progressive increase in the load on the pillar of the overlying layers. The reduction of the maximum pressure values until January 2013 was caused by the use of a pressure sensor with a lower measuring range (up to 10 MPa) at that time.

The graphs show the completely different behavior of solid and durable dolomite rocks as compared to the relatively weak sandstone rocks. Although the pressure changes in the probe vessel placed in sandstone are in the range of up to approx. 7,300 hPa, in the dolomites they do not exceed approx. 560 hPa. However, the similarity of pressure in probes on both rock mediums in terms of medium and short-term changes should be emphasized.

Definitely more susceptible to loading, the sandstone section of the pillar is characterized by several long-term trends in changes depicted in Figure 5 as subsequent phases of pillar structure destruction occurred. After the initial period of pressure drop (phase I), its long-term increase (phase II) was reached, culminating at approximately 9,200 hPa (maximum I) followed by a decisive drop (phase III) of approx. 5,500 hPa. A further period of rapid pressure increase (phase IV) occurred when the operating front was approx. 20 m from the measuring station and reached its culmination during the pillar cleavage (peak II). Phases V and VI are, respectively, the decrease and increase of pressure in the separated section of the pillar after passing the operating front.

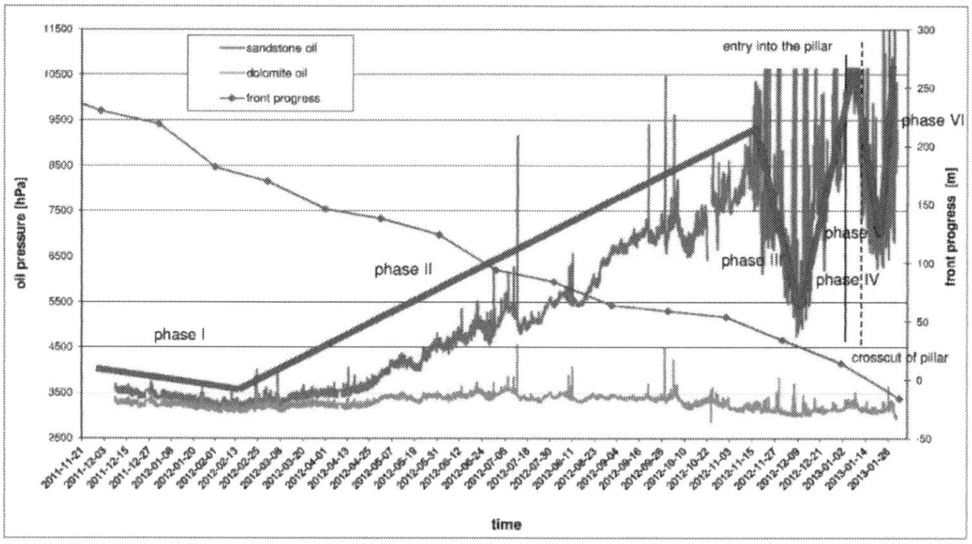

Fig. 5. Graphs of oil pressure variations measured in probe placed in the sandstone and dolomite pillar sections in relation to front progress.

Pressure changes occurring in the dolomitic pillar are discussed in the modified graphs obtained from their average (moving average, n = 50) and the stretched ordinate corresponding to the pressure in the dolomite medium (Figure 6). After this time period (from 11/2011 to 11/2012), the high compatibility of the pressure changes in the sandstone and dolomite mediums, in the dolomite section of the pillar, tended to decrease, with a minimum convergence for the sandstone section of the pillar. This decline temporarily precedes the occurrence of peak I for the sandstone medium and is roughly equivalent to the third phase of destruction of this medium. Visual observations did not show the visible destruction of this section of the pillar during this period.

A detailed analysis of the variability of recorded pressure indicates that both the long-term and the medium-term trends are mainly the effects of short-term pressure pulses, successively. These pulses have different amplitudes and durations. They map the continuous process of destroying the structure of the rock medium, clearly indicating its sequential character. This is expressed in the form of medium-term pressure (temporary waveform) changes and the long-term trend.

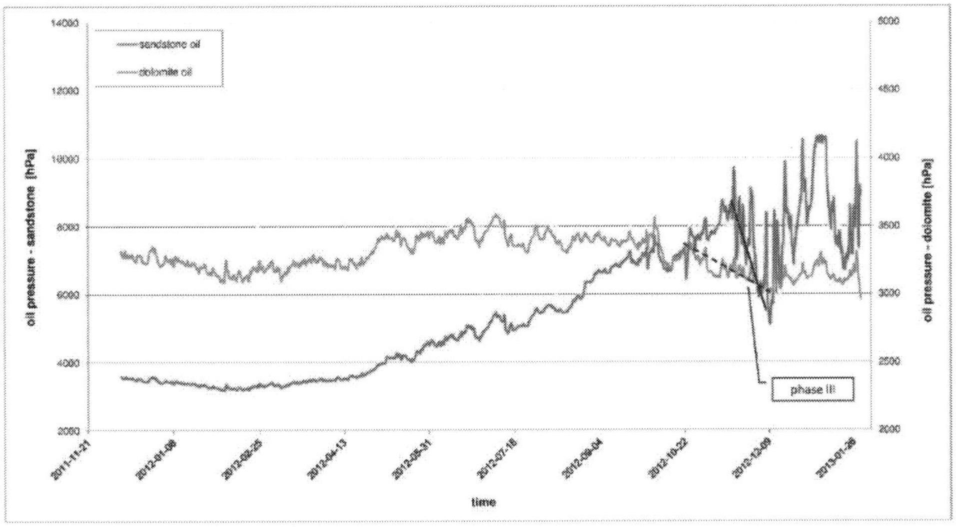

Fig. 6. Smoothed graphs of pressure changes of probe measured in the dolomite and sandstone pillar sections with the marking of the phase III of deformation.

5 Comparative analysis of the test results on rock samples and in situ conditions

The comparison of the destruction of the rock medium obtained from rock samples and actual pillar operating conditions indicates the limited similarity of their strength characteristics. In the case of the sandstone pillar sections, the compatibility of these characteristics concerns the presence of phases I and II, peak I and phase III, which must be interpreted successively as: the pre-critical condition, the moment of destruction and the post-critical condition. The fundamental difference between the analyzed objects concerns the presence of a further phase of pillar deformation with the appearance of the subsequent phases, in particular phase IV.

The presence of this phase may indicate the destruction of the pillar structure and the occurrence of the so-called strengthening the structure of the rock medium effect as a result of the reconsolidation process. Phase IV of pillar structure destruction does not occur in the case of rock samples because they are already completely destroyed. The further deformation stage (peak II, phases V and VI) is a consequence of the weakening of the pillar structure as a result of its cleavage. Earlier in relation to the sandstone medium, the third phase of deformation in the dolomite pillar reveals the material properties of this medium and its deformation. Assuming the main forces in the vertical direction and the nature of the influence of these forces on the probe, the observed drop of pressure in the probe placed in the dolomite can only be due to the reduction of lateral pressure exerted on the probe. This is probably indicative of permanent deformations of the dolomitic medium in that direction, i.e. perpendicular to the pillar. These changes are closely correlated with the response to the loading of the sandstone section of the pillar and the vulnerability of the sandstone rock mass in this area, i.e. the medium with significantly weaker strength parameters.

The second significant difference in the strength characteristics of the rock samples and the actual pillar consists in detailing the variation in the measured pressure for the pillar. Unlike a generally smooth curve describing the behavior of a rock sample subjected to a constant loading rate, a very irregular course is observed for the actual pillar in the form of the aforementioned short- and medium-term changes. The most probable source of short-term pulse pressure variations is the wide range of displacement in elemental volumes of the rock medium, ranging from rock micro-defects to large cracks. On the other hand, medium-term pressure changes should be related to temporal and local consolidation of the structure of the rock medium.

The variability of the deformation characteristics of the rock samples and the actual pillars is the result of unfavorable test conditions. It should be assumed that the most important factors that determine this are the size and proportion of the rock sample dimensions and the loading rate in the strength test machine.

6 Conclusion

The article presents the experimental results carried out under pit conditions i order to determine the rock pillar reaction to the increasing rock mass. Knowledge of the actual deformation characteristics of the structural pillars in room and pillar operating systems is of paramount importance in the course of selecting the optimum dimensions of these pillars and their response to the operation of the bed [9-11]. So far, the process of deformation of the pillars has been presented from the results of laboratory measurements on rock samples, indicating their pre-critical state, the moment of destruction and the post-critical state (post-destructive). With respect to the actual conditions for larger pillars, the moment of complete destruction of the sample corresponded to the first peak in the pressure gradient measured in the sandstone part of the pillar. According to this comparison, it should be considered that the pillar has already reached residual resilience and should not pose a threat as a result of its sudden destruction. In practice, the pillar structure is reclassified and its strength increases. In the case described above, this situation occurred when the front line was aligned with the pillar in question.

No reconsolidation effect in rock sample studies is probably due to the very small size of the sample. Therefore, in order to better represent the real working conditions of the pillars in the room and the pillar operating system, it seems reasonable to extend the strength tests to larger rock samples and to reduce the proportion of their slenderness to transverse dimensions [12]. In identifying the actual working conditions of the structural pillars should consider in the greater complexity of the models studied in the strength test machine.

References

1. J. Butra, *Eksploatacja złoża rud miedzi w warunkach zagrożenia tąpaniami i zawałami* (Wydawnictwo KGHM Cuprum, Wrocław, 2010)
2. Z. Bieniawski, Int. J. Rock. Fract. Mech. **4**, 4, 415-430 (1968) doi 10.1007/BF00186807
3. W.R. Wawersik, C. Fairhurst, *A study of brittle rock fracture in laboratory compression experiments*, Int. J. Rock. Mech. Min. **7**, 5, 561-575 (1970)
4. A. Nur, J. D. Byerlee, J. Geophys. Res. **76**, 6414-6419 (1971) doi 10.1029/JB076i026p06414
5. M. Kwaśniewski, *Dylatancja jako zwiastun zniszczenia skały. Część I. Fizykalna istota zjawiska*, Przegląd Górniczy **2**, 42-49 (1986)
6. J. Lis, P. Kijewski, *Badania własności wytrzymałościowych i odkształceniowych skał z kopalń LGOM w trójosiowym stanie naprężeń*, Czasopismo Naukowo-Techniczne Cuprum **4**, 17-40 (2007)
7. J. Lis, P. Kijewski, *Wpływ prędkości deformacji modeli skalnych na ich zachowanie w stanach przed i pozniszczeniowych*, Zeszyty Naukowe AGH, seria Górnictwo **129**, 191-197 (1987)

8. W. Grzebyk, L. Stolecki, *Pomiary zmian objętościowych masywu skalnego jako wskaźnik oceny stanu górotworu*, Materiały Konferencyjne: Górnicze Zagrożenia Naturalne, Katowice, Poland (2013)
9. R. K. Zipf, *Toward pillar design to prevent collapse of room-pillar mines*, Proc. of 108th Annual Exhibit, Society for Mining, Metallurgy and Exploration Conference SME (2011)
10. C. D. Martin, W. G. Maybee, *The strength of hard-rock pillars*, Int. J. Rock. Mech. Min. **37**, 1239-1246 (2000)
11. G.S. Esterhuzien, D.R. Dolina, J. L. Ellenberger, *Pillar strength in underground stone mines in the United Stated*, Int. J. Rock. Mech. Min. **48**, 42-50 (2011)
12. W.J. Darlington, G.R. Pathegama, S. K. Choi, Rock. Mech. Rock. Eng. **44**, 513-529 (2011) doi 10.1007/s00603-011-0161-6

Petrological features of selected components of the Cergowa sandstones (Outer Carpathians) recorded by scanning electron microscopy – preliminary study

Joanna Pszonka[1,*]

[1]Mineral and Energy Economy Research Institute, Polish Academy of Sciences, 31-261 Krakow, Poland

Abstract. The scanning electron microscope analysis of the Cergowa sandstones brings new data on their petrological features and chemical composition. Previous work in standard petrographic examination, e.g. polarising (PL) or cathodoluminescence (CL) microscopy, displayed limited information on grain surface topography and only assumptions to their geochemistry. Both identification and characterisation of minerals are fundamental in the progress of mining and minerals processing systems. Detrital grains of the Cergowa sandstones are bound by calcite and dolomitic cement and commonly corroded by diagenetic fluids, however, in varying degrees, which is illustrated here by feldspar, quartz and dolomite minerals. Dissolution processes of marginal parts of these mineral grains resulted in corrosion, which increased the contact surface between the grains and the cement. The difference in resistance to these processes was observed not only among distinct groups of minerals, but also within the group of feldspars: between K–feldspars and minerals of plagioclase. That combination resulted in exceptionally strong cementation of the Cergowa sandstones, which is expressed by their high hardness and resistance to abrasion, freezing, and thawing. Inherent parameters of sandstones are characterised by their petrographical properties.

1 Introduction

Previous studies about the Cergowa sandstones, conducted mainly by using polarising (PL) or cathodoluminescence (CL) microscopy [1–4], clearly demonstrate advantages of complementary approaches to its petrographic properties. As an example, cathodoluminescence microscopy proved to be a very usable and even indispensable method in studying the Cergowa sandstones and thereby often reveals features not distinguished by other techniques [3, 4]. Nevertheless, CL microscopy also demonstrates difficulties in obtaining some data, e.g. the geochemical composition of single detrital grains and surrounding cement [4]. Used techniques in the Cergowa sandstones petrology did not enable more detailed description of a chemical composition of rock components,

* Corresponding author: jpszonka@min–pan.krakow.pl

thereby allowed for assumptive identification of minerals. The combination of CL imaging with CL spectroscopy proved even some difficulties in the study material because of too small grains size. Furthermore, luminescence of carbonates, which constitute a large proportion of these sandstones, is brighter than other components such as silicate minerals, which in such conditions show little or no luminescence.

Benefits and weaknesses of used methods have prompted research with supplementary techniques: "Further work, including geochemical investigations (...) should be conducted in the future" [3]. Scanning electron microscope (SEM) was chosen to further investigation, because provides higher magnifications, better detection and more stable operating conditions than conventional techniques. The scanning electron microscopy enables to obtain three–dimensional images of individual sediments grains of the Cergowa sandstones and their chemical composition. Such results significantly complete foregoing research of the Cergowa sandstones petrology [1, 2, 4].

2 Material and methods

The Cergowa sandstones (Lower Oligocene) constitute the main part of the lithosome Cergowa Beds and are situated within the Dukla and Fore–Dukla tectonic unit of the Eastern Outer Carpathians [5–7]. The Cergowa sandstones represent sediments deposited by a variety of mass gravity flows as submarine fan or fans [8–11] and were classified as greywackes or lithic wackes and arenites [1, 2].

Samples of the Cergowa sandstones were collected from outcrops at five locations: four from the Dukla tectonic unit (Iwla, Lipowica, Stasiana and Wernejówka) and one from the Fore–Dukla tectonic unit (Rudawka Rymanowska). Hand specimens of sandstones for SEM examination were crushing to give pieces of suitable size (a few millimetres) for mounting. Each sample was glued to the stub (1 cm in diameter) and coated with carbon to provide conduction. In total, 10 samples were analysed. Chemical composition of minerals was detected by the energy dispersive spectroscopy (EDS), using standard analytical algorithms, where the element concentrations are normalized to 100%.

Investigation was performed at the SEM–Laboratory in the Geological Institute at the TU Bergakademie Freiberg with the usage of the Jeol JSM 7001F scanning electron microscope. The imaging was carried out in the Back Scattered Electron (BSE) Bruker, Esprit 1.9.3. with following parameters: accelerating voltage of 15 keV, electron beam current of 10 μA, and working distance 12–14 mm.

3 Results

3.1 Feldspars

SEM images of feldspars display difference between a grain surface morphology of K–feldspars and plagioclases. The grain surface topography of K–feldspars shows more intense etching and is generally rougher than that of plagioclases (Fig. 1, top). Emission bands, under the SEM image, enable to identify and distinguish both mentioned feldspars (Fig. 1, middle and bottom).

 DOI: 10.1051/e3sconf/20172402004

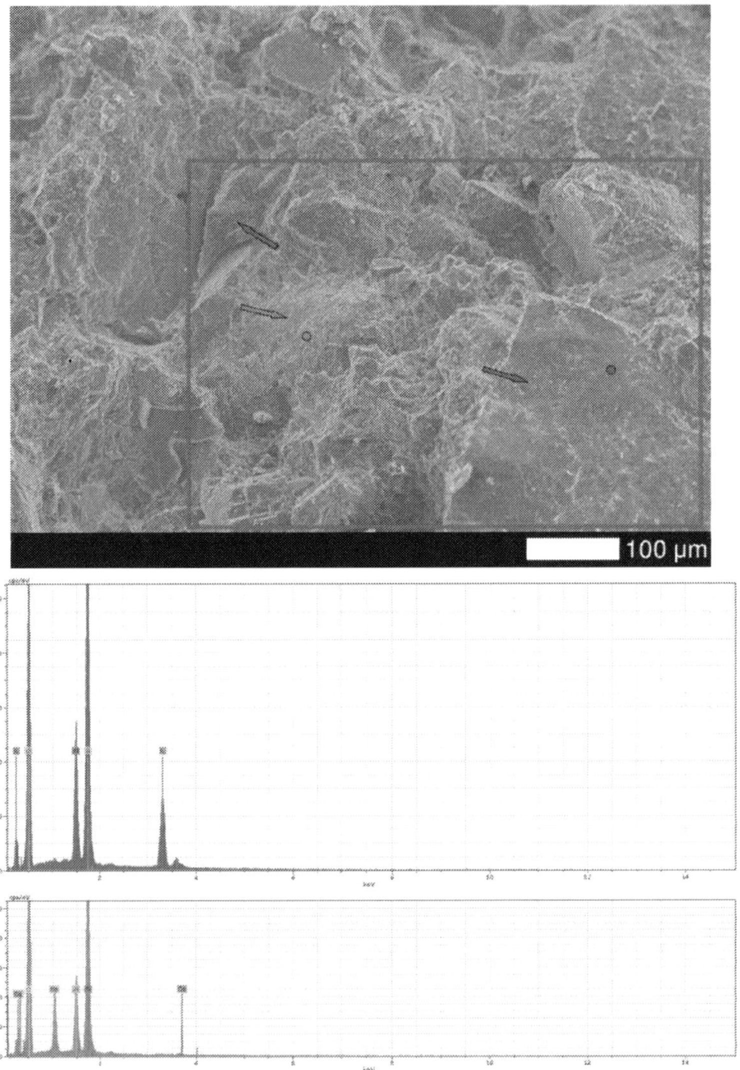

Fig. 1. Top – SEM image of feldspar minerals: blue arrow points to K–feldspar, red arrows point to minerals of plagioclase solid solution. Middle – Emission band displays chemical composition of the K–feldspar recorded in the blue spot. Bottom – Emission band displays chemical composition of the plagioclase recorded in the red spot.

3.2 Quartz

Quartz belongs to the most abundant group of minerals in the Cergowa sandstones. However, due to short–lived luminescence colours, brownish colour or no luminescence of quartz from tectonically active region, it was often overlooked in cathodoluminescence studies. Surface relief of quartz in SEM images, is marked by few trails of dissolution (Fig. 2, top and 3). Even some irregularly etched quartz grains indicate their high resistance and/or low intensity for weathering and diagenetic processes. Emission band in the figure 2 (bottom) distinguishes quartz among remaining components of the sandstone.

 DOI: 10.1051/e3sconf/20172402004

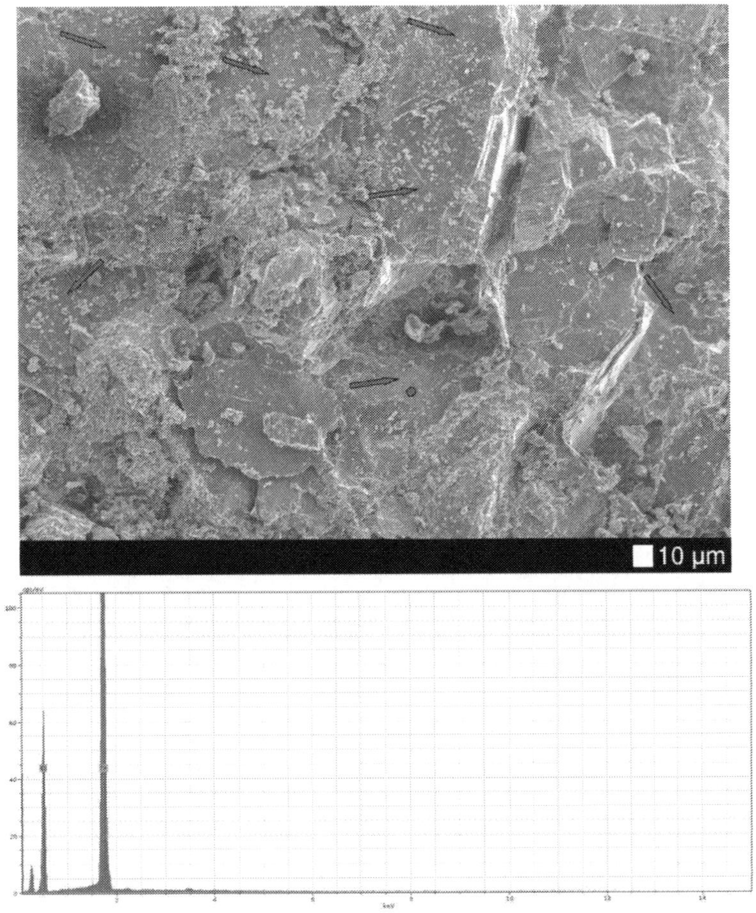

Fig. 2. Top – SEM image of quartz pointed by red arrows. Bottom – Emission band displays chemical composition of quartz recorded in the red spot.

Fig. 3. SEM image of quartz pointed by red arrows and carbonate cement pointed by blue arrows, surrounding quartz grains.

3.1 Dolomites

Dolomites in the Cergowa sandstones occur as lithic grains and cement. Cathodoluminescence imaging allowed recognising polycrystalline detrital grains of dolostone as well as monocrystals occurring as detrital components and cement [4]. However, using CL imaging without spectroscopy it was only possible to estimate the chemical composition of the Cergowa sandstones components.

The SEM visualisation has revealed finely, rhomb–shaped, densely packed, euhedral to subeuhedral dolomite crystals which constitute cement in the Cergowa sandstones (Fig. 4 and 5), and random, solitary, larger dolomite grains, which are identifiable by rhomboidal shape and etched surface morphology and represent detrital components (Fig. 4 and 6). Moreover, the relationship between calcite and dolomite cement is here clearly seen (Fig. 5), whilst in standard techniques it is obliterated [4]. The chemical composition of detrital and authigenic dolomite is illustrated by emission bands in figure 4 (bottom) and 5 (bottom).

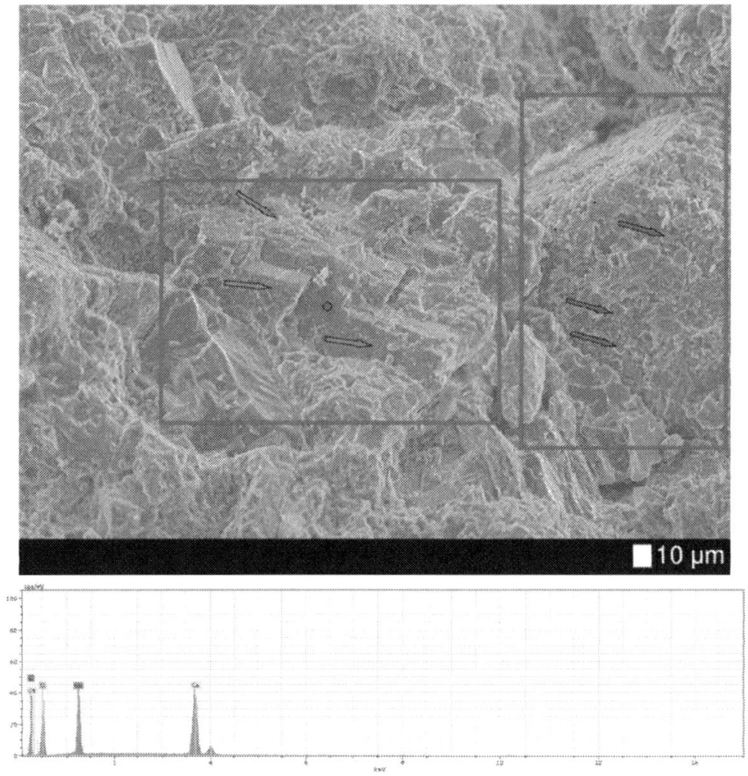

Fig. 4. Top – SEM image of randomly distributed detrital grain of dolomite pointed by green arrows and finely, densely packed, euhedral to subeuhedral dolomitic cement pointed by red arrows. Bottom – Emission band displays chemical composition of detrital dolomite recorded in the red spot.

 DOI: 10.1051/e3sconf/20172402004

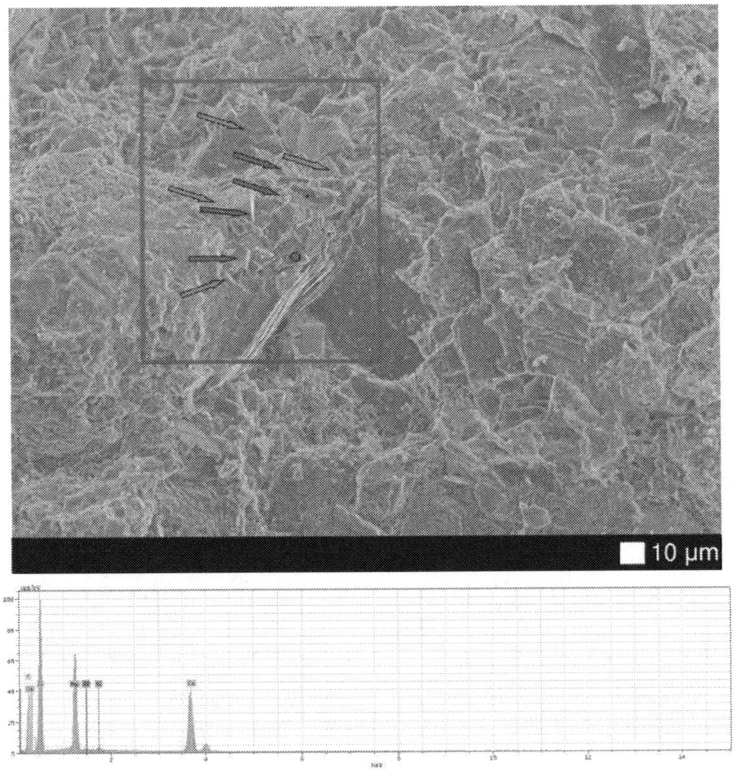

Fig. 5. Top – SEM image of finely, densely packed, euhedral to subeuhedral dolomitic cement pointed by red arrows, surrounded by calcite cement pointed by blue arrows. Bottom – Emission band displays chemical composition of dolomitic cement recorded in the red spot.

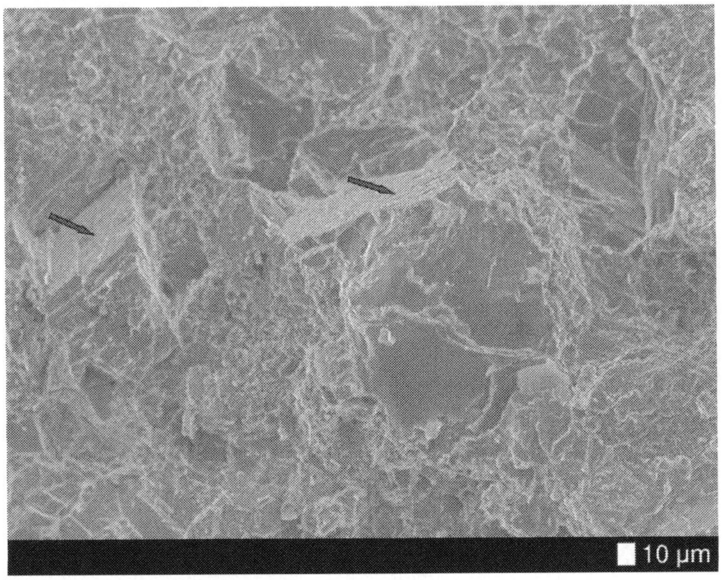

Fig. 6. SEM image of randomly distributed detrital grains of dolomite (pointed by red arrows) with partly abraded surface morphology.

4 Discussion

The etching, embayment and partial replacement of the Cergowa sandstones grains during carbonate precipitation are more and more imaged and known, simultaneously reflect conditions in which dissolution was produced. Acidic pore waters had destructive effect on all minerals of the framework, with simultaneous precipitation of carbonate material [1-4]. Growth of calcite in place of pre–existing detrital grains is very clearly visible in SEM images of the Cergowa sandstones.

The acidic environment was favourable for dissolution and destruction of detrital grains but with varying intensity. SEM results in this publication visualize this diversity, which is displayed not only among different groups of minerals such as feldspar, quartz and dolomite, but even in the field of feldspars. Obtaining such visualisation was indistinct by using polarizing (PL) and cathodoluminescence (CL) microscopes. Thereby, the CL imaging of feldspar minerals replacement by carbonate cement [3, 10] presumably relates to K–feldspars and not to the whole group of feldspars as reported in the cited publications. Besides low and moderate stage in etching of quartz and dolomite verify results of the previous works in standard petrographic techniques in regard to low intensity of dissolution [1].

Despite its modest intensity, this process was very important for the sediment lithification. It resulted in increased surface of the corroded grains, and therefore stronger cementation of the components, which is manifested by high hardness and resistance of the Cergowa sandstones [12, 13].

5 Conclusions

1. The SEM analyses enabled to document new petrological features of the Cergowa sandstones, which the PL and CL observations do not reveal, such as:
   - chemical composition of components,
   - difference between a surface relief of K–feldspars and plagioclases,
   - sporadically and irregularly etched quartz grains indicating their high resistance,
   - visualisation of variety in shapes and size of dolomite.
2. Dissolution was important for the sediment lithification by increasing the surface of contact between corroded grains and carbonate cement; this resulted in very strong cementation of the Cergowa sandstones.
3. This study shows that the SEM analysis and the standard petrographic techniques, as PL and CL microscopy, should be considered as complementary techniques.

Many thanks to Anja Obst (TU Bergakademie Freiberg) who helped with the scanning electron microscope analyses, which were funded by the Post–Doctoral Research Grant of the International Association of Sedimentologists 2016/2017 for PhD Joanna Pszonka.

References

1. C. Peszat, *Variations of mineral composition of the Cergowa sandstones in the light of their deposition conditions and diagenetic alterations* (in Polish with English abstract), Biuletyn Państwowego Instytutu Geologicznego, **346**, 24, 207–234 (1984)
2. J. Pszonka, *The influence of sedimentation and diagenetic processes on economic significance of the Cergowa sandstones from "Lipowica II–1" deposit*, Gospod. Surowcami Min. **25**, 3, 333–342 (2009)

3. J. Pszonka, M. Wendorff, Gospod. Surowcami Min. **30**, 4, 21–36 (2014) doi: 10.2478/gospo–2014–0036
4. J. Pszonka, M. Wendorff, Int. J. Earth Sci. **106**, 269–282 (2017) doi: 10.1007/s00531–016–1318–z
5. M. Cieszkowski, A. Ślączka, W. Zuchiewicz, *Detailed geological map of Poland 1:50000, Jaśliska sheet* (PGI, Warszawa, 1990)
6. A. Ślączka, *On the possibility of occurrence of bitumen deposits in the western part of the Dukla Unit in the Polish Eastern Carpathians* (in Polish with English abstract), Geol. Q. **12**(2): 344–249 (1970)
7. A. Ślączka, *The Geology of the Dukla Unit, Polish Flysch Carpathians* (PGI, Warszawa, 1971)
8. D. Dirnerová, M. Prekopová, J. Janocko, Geol. Q. **56**, 3, 547–560 (2012) doi: 10.7306/gq.1039
9. P.S. Dziadzio, Z. Borys, S. Kuk, E. Masłowski, J. Probulski, M. Pietrusiak, A. Górka, J. Moryc, P. Karnkowski, P.H. Karnkowski, M. Pietrusiak, *Hydrocarbon resources of the Polish Outer Carpathians – reservoir parameters, trap types and selected hydrocarbon fields: A. Stratigraphic Review* (AAPG, Tulsa, 2006)
10. J. Pszonka, *Studium sedymentologiczne warstw cergowskich w jednostce dukielskiej i przeddukielskiej Karpat fliszowych* (IGSMiE PAN, Krakow, 2015)
11. A. Ślączka, R. Unrug, *Trends of textural and structural variation in turbidite sandstones: the Cergowa Sandstone (Oligocene, Outer Carpathians)*, Ann. Soc. Geol. Pol. **46**, 55–76 (1976)
12. C. Peszat, M. Buczek–Pułka, *Znaczenie surowcowe piaskowców cergowskich na tle ich właściwości fizyczno–mechanicznych*, Biuletyn Państwowego Instytutu Geologicznego, **351**, 179–211 (1986)
13. J. Pszonka, M. Wendorff, E. Kusaj, *Piaskowce. cergowskie jako najważniejszy surowiec skalny województwa podkarpackiego*, Biuletyn Państwowego Instytutu Geologicznego, **439**(2), 261–268 (2010)

Use of petrophysical data for siting of deep geological repository of radioactive waste

Liliana Petrenko[1,*], and *Vyacheslav* Shestopalov[1]

[1]Institute of Geological Sciences of NAS of Ukraine, 01054 Kyiv, Ukraine

Abstract. The paper is devoted to analyzing the petrophysical properties and petrographical characteristics of Volyn region with the view to choosing the least permeable and so the most suitable geological formation for the radioactive waste disposal. On a basis of the petrophysical estimations of the granitoids properties the argumentation of permeability has been developed for the petrotypes of Volyn region. Also method of classification of the petrotypes with their relative rate of suitability for radioactive waste disposal was developed. As a result of studying the perspectives were shown of the zhytomyr and korosten types of the granitoids as host rock for the radioactive waste disposal. According to the results of investigations performed by Swedish researchers a comparative analysis of rocks based on the age of formation, composition, structural features and some petrophysical properties of granitoids as host rocks for repository of radioactive waste was performed. Detail comparison the data of the granitoids of the Forsmark site in Sweden and the data of the granitoids of the Volyn megablock can be one of the next steps in researching the host rocks for the development of the RW disposal system in Ukraine.

1 Introduction

In accordance with the international experience of radioactive waste management, geological repository designed for a long-term use, is the only technically feasible alternative to safe isolation of the most dangerous radioactive waste (RW). In such a system rocks play an important role.

According to the State Program [1] and the Strategy for Radioactive Waste Management [2], the geological repository of RW should be located within the Chornobyl Exclusion Zone (CEZ) or the adjacent areas. This approach provides the greatest safety for radioactive waste transfer to their place of disposal. Because of this, the role of the host geological environment for the disposal of radioactive waste is preferred by the granitoid formations of the Volyn megablock and its surroundings.

Assessment of the "internal" permeability as the permeability of rocks around the disposal facility, and not only the "transit" characteristic of the environment is an important

* Corresponding author: Petrenko.L@nas.gov.ua

task for siting of safe geological repository of radioactive waste. The problem is that today most of the existing remote methods (seismic and other geophysical studies, as well as, in part, aerospace method) for direct determination of the filtration properties of deep-lying rocks in their natural occurrence are aimed at obtaining information on the relative permeability of the rock blocks, the bodies, and not of specific geological formation in the zone of waste packages disposal. Thus, the application of the above studies is not sufficiently complete.

Evaluation of permeability on the basis of analysis of petrophysical properties (density, porosity, thermal conductivity, etc.) and petrographic peculiarities will not allow us to directly determine the rock permeability, but only indirectly. It is equally important that the obtained results of the analysis of the petrophysical characteristics of granitoids for the Volyn megablock can be extrapolated to the CEZ rocks prior to the prospecting and exploration work stages. This will allow us to focus on more suitable areas.

Data for researches provided by the Problem Research Laboratory of Physical and Chemical Researches of Rocks of the Institute of Geology National Taras Shevchenko University of Kyiv.

2 Results

2.1 The structure and texture features of the main complexes of the Volyn megablock

According to Swedish System [3], the initial stage of siting the repository of radioactive waste is to select sites that are characterized by geological homogeneous composition and structure of rocks. One of the criteria for selecting a site at the stage of regional research is the following: the site should be located within the watershed areas, outside the rivers floodplains, out of the regional fault zones, within the areas of low density of local faults, within the areas of low gradient geophysical fields.

The Volyn megablock is located in the northwest part of the Ukrainian Shield (US). It is composed of the deposits of the sedimentary cover (formations of the Mesozoic and Cenozoic systems) and the Archaean-Proterozoic crystalline rocks of the basement. Weathering crust is of mosaic occurrence. The structure of the Volyn megablock includes seven complexes (Perzhanskyi, Korostenskyi, Kyshynskyi, Osnytskyi, Bukynskyi and Zhytomyr) and 35 key petrotypes. Petrotype must be understood as the geologically similar type, formed in specific and rather limited thermodynamic, physical and chemical conditions, characterized by comparable composition, structural and texture features and physical properties [4]. Below and thereafter the main attention is paid to the complexes and their petrotypes, which are most commonly developed in the CEZ. Zhytomyr complex is represented mainly by uniformly granular, occasionally porphyry-shaped granites with massive texture. In the anticlinal structures granites form separate stranded blocks of medium-grained uniformly granular muscovite-biotite granites up to 20 km^2 in area (Zhytomyrskyi etc), small massifs of porphyry-shaped granites, as well as larger massifs of porphyry-shaped granites. The largest massifs - Korostyshivskyi, Kamianobrodskyi, Fedorivskyi, and Krasnogorskyi have the area of 100-300 km^2. Granitoids of the Korostenskyi complex, together with the magmatites of basic (subsilicic) composition (gabbro, anorthosites) form the Korostenskyi (Korosten) Pluton. The petrotypes of the Korosten complex belong to the medium-and coarse-grained fraction of the massive texture, except for the Ignatpilskyi rapakivi-like granite, which is finely and medium-grained.

2.2 Petrophysical characteristics of the main petrotypes of the Korostenskyi and Zhytomyr complexes

The purpose of the petrophysical assessment is based on the separation of granitoids of the Volyn megablock into relatively more suitable and relatively less suitable ones in accordance with the requirements for rock formations when creating a waste disposal system. The physical properties of crystalline rocks, which characterize their minimal permeability, correspond most to the conditions of safe disposal of RW. They involve porosity and density. Thermal, magnetic and strength characteristics of rocks are closely related to them. These properties (see the average value of the petrophysical characteristics of the key petrotypes in Table 1), as well as rock composition, structural and textural features should be taken into account when substantiating the suitability of the petrotype as the host rock environment for creation of a RW disposal facility.

Table 1. The average value of the petrophysical characteristics of the key petrotypes of the Korostenskyi and Zhytomyr complexes.

Petrotypes	Porosity (open), n_o, %	Porosity (general), n_g, %	Density (volumetric), σ, kg/m^3	Thermal conductivity, λ, W/m·K	Coefficient of thermal heterogeneity, β	Young's module, E,×10^{10}	Compression module, K, ×10^{10}	Shear modulus, G, ×10^{10}
Korostensky complex								
granite Liznikovsky	0.59	1.79	2617	3.49	0.14	7.0	4.1	2.9
rapakivi-like granite Ignatpilsky	0.12	1.14	2630	3.27	0.13	7.1	4.5	2.9
rapakivi-like granite Norinsky	0.37	2.48	2595	2.80	0.17	6.5	4.5	2.6
rapakivi-like granite Rozsokhivsky	0.37	1.15	2601	3.16	0.24	6.9	4.6	2.7
granite Berezovsky	0.33	2.10	2612	3.39	0.22	7.1	4.5	2.9
rapakivi-like granite Emelyanovsky	0.32	1.64	2621	3.11	0.31	6.0	3.5	2.5
rapakivi-like granite Korostensky	0.40	1.73	2618	3.21	0.14	6.6	3.7	2.8
rapakivi-like granite Potyivsky	0.26	2.18	2659	2.94	0.30	8.1	5.2	3.3
granite rapakivi- Malinsky	0.31	1.66	2635	2.92	0.17	6.9	4.3	2.8
Zhytomyr complex								
granodiorite Kornynsky	0.17	1.90	2707	-	-	-	7.9	-
granite Korostyshevsky	0.36	2.30	2644	2.56	0.17	6.3	3.5	2.6
granite Zhytomyr	0.22	0.87	2658	2.77	0.06	7.3	4.8	2.9
granite Kurchytsky,	0.30	1.59	2661	3.02	0.16	7.0	4.2	2.8
granite Kamianobrodsky	0.23	1.46	2653	3.22	0.21	7.3	4.8	2.9
Granodiorite Olexandrivsky	0.25	1.70	2699	3.00	-	8.5	4.8	3.2
Note *: the data range is ± 3-5%								

The basis for assessing the petrophysical properties of rocks as the formation for radioactive waste disposal is the substantiation of the numerical values of the signs of suitability and selection of the reference petrotype, according to these features. This allowed us to classify all the petrotypes into conditionally suitable and conditionally unsuitable ones relatively to the reference petrotype.

The analysis of the suitability of the key petrotypes by a particular petrophysical feature does not seem expedient, since the characterization of permeability as the main property of a crystalline rock in the context of creating safe RW repository is of complex character. In order to simplify the consideration of petrophysical properties, they were divided into groups: permeability properties (general and open porosity), density properties (mineral and volume density), thermal properties (thermal conductivity, coefficient of thermal inhomogeneity), magnetic properties (magnetic susceptibility, natural magnetization intensity, Kenigsberg factor).

2.2.1 Group of properties characterizing permeability

Porosity and permeability are important physical characteristics of rocks when choosing a suitable geological formation for the disposal of radioactive waste. The void through which the aqueous solutions can migrate also affects the strength of rocks. Thus, rocks of different mineral content can have the same values of strength due to different values of porosity and permeability. It was established that permeability increases with increasing porosity. However, for different types of rocks, permeability for the same porosity is different. At equal porosity, the more homogeneous rocks will be characterized by minimum permeability, and rocks with relatively large and equal in size pores - by maximum permeability [5].

2.2.2 Group of density properties

The main factors determining the rock density are the mineral and chemical composition of the main rock-forming minerals, structural and textural features of rocks, their degree of diagenesis and metamorphism. In addition, the density of rocks is associated with their position in different tectonic structures, with the age of their formation, regional and local changes in area and in vertical. The density of the igneous rocks depends mainly on the rock composition and grows with increasing their basicity. Porosity has a noticeable effect on density: relatively low density values are proved by increased values of porosity.

2.2.3 A group of thermal properties

At the time of RW disposal, the release of radiogenic heat will continue. Accordingly, rocks should have sufficient heat conductivity and heat capacity, which will promote its scattering in the environment. An example of petrotypes of the Volyn megablock was the analysis of the dependence of heat conductivity on the mineral composition and grain size, total and open porosity. In particular, the high content of quartz and SiO_2 corresponds to a higher value of thermal conductivity. Dependence of the latter on the size of the minerals' grains has directly proportional character: with decreasing grain size, the thermal conductivity decreases. It also has an uneven character with respect to the rock sample, as evidenced by the coefficient of thermal inhomogeneity: its increase leads to a decrease in thermal conductivity and serves as evidence of secondary changes in the rock (metasomatism, development of fragile deformations, etc.). Thermal conductivity increase leads to a change in the volume of the rock, and this, in turn, leads to a change in strength characteristics. So, with porosity increasing thermal conductivity decreases. The optimal solution when selecting the suitable rocks for radioactive waste disposal should be based on a comprehensive assessment of such characteristics.

2.2.4 Group of magnetic properties

Magnetic characteristics do not directly affect the safety functions of the components of the geological formation. However, their study may provide additional information on the degree of iron oxidation and its preservation. The index of oxidation to a certain extent characterizes the depth of rocks formation, which was taken into account when performing a complex petrophysical assessment.

2.2.5 Group of strength and elastic properties

The safety of radioactive waste disposal depends to a large extent on the strength characteristics of rocks. They depend on the structure of the rock, its porosity, granularity, fracturing [6]. The fine-grained rock structure is characterized by better strength than medium- or coarse-grained structure. The porphyry structure usually increases the strength of the rocks, especially when the inclusions are insignificant and the bulk of rock is fine-crystalline and monolithic. With increasing temperature (from room temperature to 200-280^{0}C), the strength characteristics increase due to increased strength of adhesion between minerals and reducing porosity. Because of low porosity, the magmatic rocks belong to the most elastic rocks.

To evaluate the elastic properties, we use indicators such as the Young's modulus E, Poisson ratio ν, total compression modulus K, shear modulus G. The module of elasticity, shear and total compression increase sharply with increasing of rock density, which is higher at the low content of silicic acid, as its specific weight is negligible. The values of the Poisson ratio give indication of the approximate quartz content. Its presence in the rocks always leads to a decrease in the ν value. Granite is characterized by the highest compressibility among the igneous rocks. Consequently, the higher values of elasticity, compression and shear modules, the less their porosity and the greater density - respectively, the rocks are characterized by greater strength.

2.3 Selection of a reference petrotype

The classification of the key petrotypes of the Volyn megablock by the degree of their relative suitability was performed aimed at the subsequent use for the disposal of radioactive waste. This classification was based on the analysis of informative properties (volume density, total and open porosity, thermal conductivity, grain size, etc.), which directly or indirectly determine the permeability of the rocks.

The relative degree of petrotype suitability can be determined by the proximity of its informative characteristics to the characteristics of a particular "reference" petrotype. "Reference" petrotype is a factitious rock whose properties were formed from the best indicators of investigated rock petrotypes most relevant to the requirements of the geological disposal safety. According to the characteristics of such a "reference" petrotype, all key petrotypes of the Volyn megablock were systematized.

Taking into account the volume of the whole sample for granitoid properties, in order to evaluate subsequent rocks selection the mode is taken as the main statistical characteristic of rocks.

Mode as one of the statistical characteristics of the sample, is an indicator of the most repetitive parameter and can be used to select the criterion signs of the reference petrotype. Estimation of the petrotype for each of the properties will be carried out in accordance with the "quality" of characteristic, and depending on its compliance with the acceptability the numerical value will be assumed being greater or lower than the mode for given petrophysical property under consideration. In the case of porosity, for example, the reference characteristic of this rock will consist of the values lower than or equal to the sample mode. Thus, the criterion values of the petrophysical characteristics were selected in accordance with the numerical value of the mode of each property.

Considering the volume density and the heat conductivity, the "suitable" petrotype by given feature can be petrotypes with a numerical value greater than the criterion (numerical value of the mode). In the case of general and open porosities petrotypes with a numerical value lower than the reference can be attributed to more suitable ones.

Thus, to determine the petrotype standard for granites of the Volynsky megablock, the petrophysical properties corresponding to the mode of sample for each property were chosen into the criterion values of the signs (taking into account the class according to the rock structure). The reference prototype signs are shown in Table 2.

Table 2. Signs of the reference petrotype.

Porosity (open), n_o, %	**Porosity (general), n_g, %**	**Density (volumetric), σ, kg/m^3**	**Grain size (G.s.), mm (class)**	**Coefficient of thermal heterogeneity, β**	**Thermal conductivity, λ, W/m·K**	**Young's module, E,×10^{10}**	**Compression module, K, ×10^{10}**	**Shear modulus, G, ×10^{10}**
less 0.3	less 2.0	more 2630	$10^{-1} - 2.5\times10^{-1}$ (2)	less 0.17	more 3.1	more 7.1×10^{10}	more 4.8×10^{10}	more 2.9×10^{10}

When analyzing the petrotypes of the Volyn megablock relative to reference petrotype, they are placed in a certain hierarchy by petrophysical properties. In order to further classify them by the degree of suitability, the "weighting" factor of petrophysical properties was taken into account when choosing a safe geological formation for radioactive waste disposal.

The values of weighting factor were determined according to the contribution to the safety of the disposal facility in the conditions of its placement in crystalline rocks. In particular, in these conditions, the highest coefficient is given to the index of rock characteristic that is most closely related to permeability. Thus, International Atomic Energy Agency (IAEA) recommendations were taken into account, according to which the information on the porous and fractured rock structure is important for ensuring the long-term performance of the repository system. The thermal and thermo-mechanical properties of rocks were taken into account as well.

After comparing each of the key petrotypes of the Volyn megablock with the reference, they will be characterized by the coincidence of a certain number of signs. The use of the transition coefficient allowed us to estimate the "weight" load of these characteristics and to determine the relative degree of the petrotypes suitability for the disposal of radioactive waste. Table 3 summarizes the significance of petrophysical characteristics and their compliance with safety requirements.

Table 4 shows the degree of proximity of the some key petrotypes of the Korostenkyi and Zhytomyr complexes to the reference petrotype, taking into account the weight coefficients of the characteristics used. According to Table 4, such petrotypes as Ignatpilskyi granite, Zhytomyr granite and Kamyanoborodskyi granite are close to the reference petrotype by the greatest number of signs.

Based on the results of petrophysical analysis of the key petrotypes for the Volyn megablock and taking into account the relative degrees of suitability of the petrotypes of those complexes that can be found within the CEZ and adjacent areas, the granitoids of the

Korosten complex can be provisionally considered as the most promising. The granitoids of the Zhytomyr complex may also be suitable for further research.

Table 3. Petrophysical characteristics and their compliance with safety requirements.

№	Characteristics and Value Coefficient / Reasoning	
1	Porosity (open), n_o, %	3
	It is crucial for the migration of radionuclides	
2	Porosity (general), n_g, %	2
	Significantly affects the processes of migration and fixation of radio nuclides	
3	Density (volumetric), σ, kg / m³	2
	The size of the grains largely determines the strength and thermal characteristics of the petrotype, as well as the potential predisposition of the brittle deformations, which is directly related to safety requirements	
4	Grain size (G.s.), mm (class)	2
	Evidence of the presence of secondary rock transformations or the degree of their heterogeneity, which is clearly undesirable from the standpoint of safety	
5	Coefficient of thermal heterogeneity, β	1
	Although the sign is directly related to the safety requirements, the low thermal conductivity of the enclosing rocks may be compensated by design decisions	
6	Thermal conductivity, W / m· K	1
	Has a small range of variations and is indirectly related to security requirements	
7-9	Young's module E, Compression module K, Shear modulus G, Pa	0,5
	Granites are quite strong rocks. The range of strength characteristics is negligible, so these characteristics would have an impact on the results of the evaluation	

Table 4. The degree of proximity of the some key petrotypes of the Korostenkyi and Zhytomyr complexes to the reference petrotype.

Petrotypes	n_o %	n_g, %	G.s., **mm**	**σ, kg/m³**	**λ, W/м·K**	**β**	**E,×10¹⁰**	**K,×10¹⁰**	**G,×10¹⁰**	**The total score**
Reference	3	2	2	1	1	2	0.5	0.5	0.5	12.5
Korostensky complex, granite, rapakivi-like granite										
Ignatpilsky	3	2	2	-	-	2	0.5	0.5	-	10
Rozsokhivsky	-	2	-	-	1	-	-	-	-	3
Berezovsky	3	2	-	-	-	-	0.5	0.5	-	6
Korostensky	-	2	-	-	-	2	-	-	-	4
Zhytomyr complex, granite										
Zhytomyr	3	2	-	1	-	2	0.5	0.5	0.5	9.5
Kamianobrodsky	3	2	-	1	1	-	0.5	0.5	0.5	8.5
Novoromanivsky	-	2	-	1	-	-	-	-	-	3

The petrophysical properties of granitoid of the Forsmark have been studied in sufficient detail [7, 8, 9]. The mean value of the porosity (open) for granite of the Forsmark is 0.5-0.55 %, density – 2642 kg/m³, Poisson's ratio – 0.24 [7]. Pre-comparison these values with the values of the granitoids of the Volyn megablock (the mean value of the porosity (open) for granite is 0.34 %, density – 2645 kg/m³, Poisson's ratio – 0.23) show the similarity of the granitoids of the Volyn megablock to the granitoids of the Forsmark site in Sweden. But it requires a detailed description of data from Sweden.

In the future, during prospecting and exploration works for siting of RW geological repository, the petrotypes studied may be prospective analogues and, respectively, search indicators of the host environment for development of the RW disposal system.

3 Conclusions

The paper is devoted to studying the permeability of Volyn region's granitoids with analyzing their petrophysical properties, some petrographical characteristics with the view to choosing the least permeable and so the most suitable geological formation for the RW disposal.

On a basis of the petrophysical estimations of the granitoids properties the argumentation of permeability has been developed for the petrotypes of Volyn region's as a geological formation for RW disposal has been treated. On a basis of the integrated petrophysical, structural and textural estimations of the granitoid's properties the method of the classification of the petrotypes with their relative rate of suitability for RW disposal.

The classification of petrotypes and typification of the granitoids has a perspective for further researches of these granitoids as host rock of radioactive waste disposal. As a result of studying the perspectives were shown of the Ignatpilskyi and Zhytomyr petrotype of the Korostenskyi and Zhytomyr complexes of the granitoids as host rock for the RW disposal.

The final proof of the suitability of the studied granitoids can only be obtained by the results of prospecting and exploration works in the course of safety analysis of the geological disposal system. The results obtained should be taken into account when planning and performing prospecting and exploration works that will be conducted for siting of geological repository of radioactive waste.

References

1. *National target program for radioactive waste management* http://zakon5.rada.gov.ua/laws/show/516-17
2. *On approval of the strategy for radioactive waste management in Ukraine* http://zakon3.rada.gov.ua/laws/show/990-2009-%D1%80
3. T. McEwen, T. Aikas, *The site selection process for a spent fuel repository in Finland, - summary report* (Posiva OY, Finland, 2000)
4. M.I. Tolstoy, Yu.L. Hasanov, N.V. Kostenko, A.P. Gozhyk, O.V. Shabatura, *Petrochemistry and petrophysics of Ukrainian shield granitoids* (K. Kyiv University, 2003)
5. N.B. Dortman, *Rocks and minerals* (M. Nedra, V.1, 1992)
6. V.V. Rzhevsky, G.Ya. Novik, *Fundamentals of rock physics* (M. Nedra, 1964)
7. H. Isaksson, H. Mattsson, H. Thunehed, M. Keisu, *Interpretation of petrophysical surface data. Stage 1 (2002). Forsmark site investigation.* (Report P-03-102, 2004) http://www.skb.com/publication/21314/P-03-102webb.pdf
8. J. Petersson, J. Berglund, P. Danielsson, A. Wängnerud, E.-L. Tullborg, H. Mattsson, H. Thunehed, H. Isaksson, H. Lindroos, *Petrography, geochemistry, petrophysics and fracture mineralogy of boreholes KFM01A, KFM02A and KFM03A+B. Forsmark site investigation* (Report P-04-103, 2004) http://www.skb.com/publication/22226/P-04-103webb.pdf
9. K. Tiensuu, E. Heikkinen, *KBS-3H. Geophysical logging of borehole K08028F01* (Report P-15-15, 2016) http://www.skb.com/publication/2487738/P-15-15.pdf

Preliminary studies of the dynamic stiffness modules of soil samples from the Solvay Sodium Plant waste landfill in Krakow

Elżbieta Pilecka[1,*], *Jakub* Zięba[1]

[1]Politechnika Krakowska, Instytut Mechaniki Budowli, 31-155 Kraków, Polska

Abstract. The article presents the results of laboratory tests for determining the dynamic modules of the elasticity M and the shear G, for soil samples from the landfill of the closed Solvay Sodium Plant in Krakow. The tests were performed using a triaxial apparatus equipped with "bender" piezoelements. The samples subjected to these tests were taken from two boreholes, located in the area known as the "white seas", whose formation is the result of Solvay Plant activity throughout the 20th century. The location of the test holes was planned at the place in which a road known as the" Łagiewnicka route " was planned. Studies on soil stiffness were also conducted as part of the one of the dissertation from 2008 to 2010 in the Cracow University of Technology. The results of these tests and the results of the laboratory tests that are presented in the article will be used in the designing of a computer model. This model is intended to help in assessing the dynamic impact of motor vehicle traffic on the planned Łagiewnicka route on the structure of the existing buildings located in the former Solvay Plant.

1 Introduction

During 2008-2011 numerous studies were conducted for the construction of the John Paul II Centre in Krakow to determine geotechnical conditions and construction substrate stiffness in the "white seas" waste landfill area of the former Solvay Sodium Plant in Krakow-Łagiewniki. The purpose of this study was to determine the pile carrying capacity in the anthropogenic soil, which until now, according to the norm PN-83 / B-02482, could not have been used for a structure's foundation. The following studies were conducted by several research units, including the Cracow University of Technology, Polish Academy of Sciences, and other specialist firms such as Geoteko from Warsaw. The results of the tests were finally verified through initial execution and then by testing the bearing capacity of 9 test piles buried in the ground. The results of these studies were the subject of the dissertation, "Analysis of foundation piles carrying capacity based on field surveys made by static probe" [2]. Due to the planned road construction, which was partially located in the "white seas" area of Łagiewnicka route, a typical behaviour was observed in the substrate stiffness conditions. The studies connected with the determination of dynamic parameters,

* Corresponding author: epilecka@pk.edu.pl

and the archival research of the foundation of the John Paul II Centre in Łagiewniki will allow the building of a numerical model, which will be used to analyse the impact of traffic load on the existing buildings in the John Paul II Centre in Cracow.

2 Location and geological engineering conditions

The soil samples that were studied were taken with the permission of the John Paul II Centre. The Sampling points were marked on the satellite image, where the course of Łagiewniki route is shown (fig. 1).

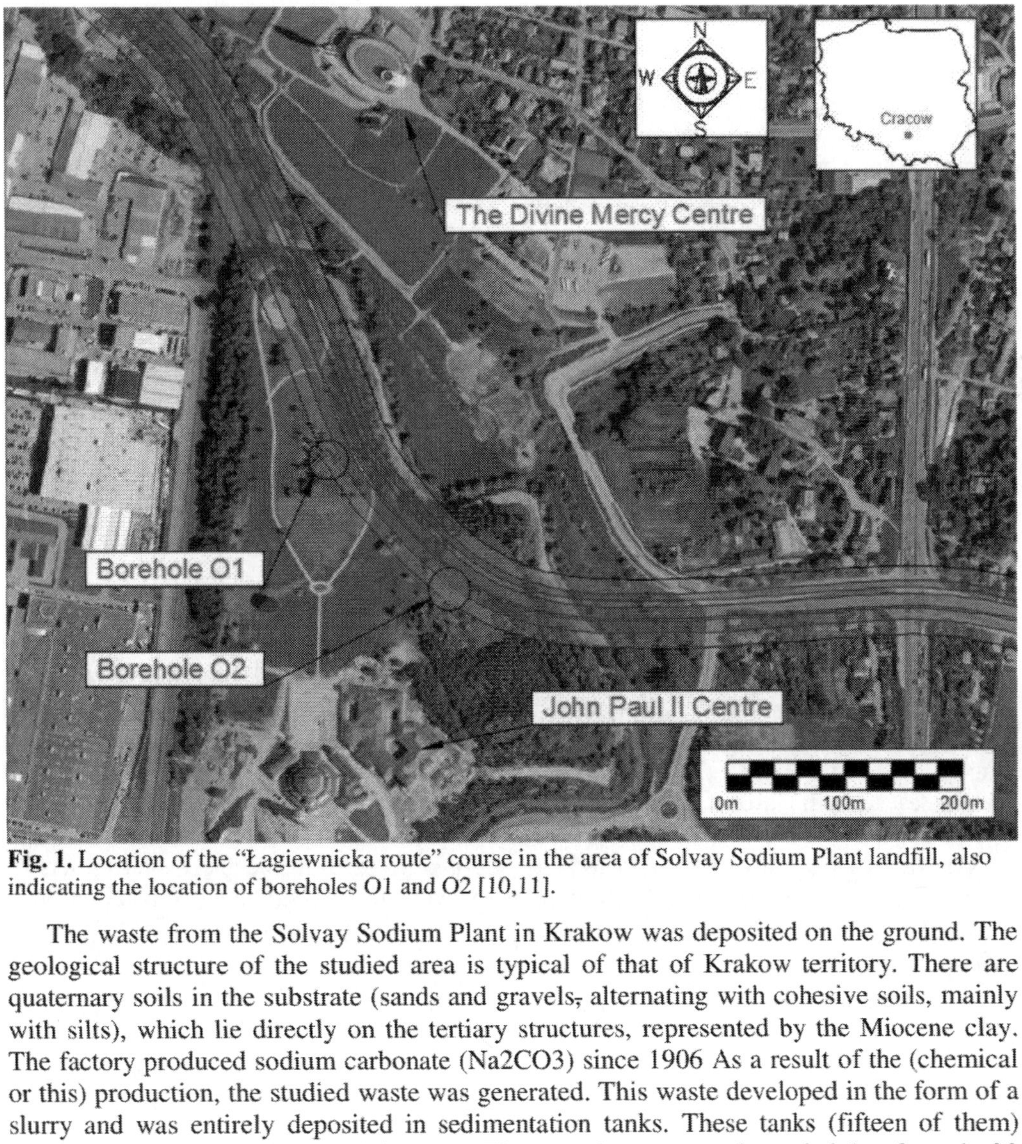

Fig. 1. Location of the "Łagiewnicka route" course in the area of Solvay Sodium Plant landfill, also indicating the location of boreholes O1 and O2 [10,11].

The waste from the Solvay Sodium Plant in Krakow was deposited on the ground. The geological structure of the studied area is typical of that of Krakow territory. There are quaternary soils in the substrate (sands and gravels, alternating with cohesive soils, mainly with silts), which lie directly on the tertiary structures, represented by the Miocene clay. The factory produced sodium carbonate (Na2CO3) since 1906 As a result of the (chemical or this) production, the studied waste was generated. This waste developed in the form of a slurry and was entirely deposited in sedimentation tanks. These tanks (fifteen of them) occupied a total area of over 30 hectares. The stored waste stood at a height of nearly 20 meters.

In the 1990s, on the basis of hydrogeological and hydrotechnical studies, the environmental effect of the waste from the Solvay Plants was analysed [3], and then the reclamation of this area was decided. The designated plan [4] and further development plans in the area of Łagiewniki forced the necessity of analysing the soil substrate and evaluating its suitability for construction purposes.

According to the plan of the Łagiewnicka route in Krakow [5], three geotechnical boreholes were made, which confirmed the occurrence of anthropogenic soils in the road subgrade. It has been found that the surface layer consists of production waste in a different state and degree of consolidation. On the basis of macroscopic examination of the test holes, the presence of calcium carbonate, sodium carbonate, and calcium sulphate was confirmed.

3 Material and methodology

The soil for the laboratory tests was taken from the two test boreholes, O1 and O2. In both boreholes below the gravel layer, the presence of anthropogenic soils was observed, mainly in the state between a semi-compact and a plastic one. The main element of the anthropogenic soil is calcium carbonate (CaCO3), which occurs together with silicon oxides (SiSO2) and other substances (i.e. P2O5, CaSO4, MgSO4, BaSO4, NaCl, CaCl2 [1]).

No water levels have been found in the boreholes. The water content in the samples varied and ranged in different degrees of saturation . Moreover, locally, there are soft plastic lime mixtures with gravel and slag with thickened (crystalized and solidified) lime. An observed characteristic phenomenon was the increased hardness of the surface layer (0.5-1.2m), composed mainly of dried, hard lime.

After being transported to the laboratory, the soil was stored at a temperature of 5° C until sample formation and testing. Due to the high variability of soil moisture in test borehole O1, samples with a natural water content measurement were taken and molded after being completely dried. The dry samples were condensed to constant volume density of all tested samples. The compaction was also due to the fact that the sample after drying revealed the porous structure and simultaneously the dried sample was crushed.

The tests were performed using a triaxial apparatus, which is equipped with a wave generator and caps attached to the tested sample. This apparatus holds the ability to generate S-wave and P-wave. The waves are then excited in a piezoelectric bender elements device, and they deform due to a change in tension. Using the triaxial apparatus, velocity wave measurements were made during isotropic tension. The soil strain corresponds to the range of $10^{-4} \div 10^{-5}\%$ [6].

The test equipment used in this study is shown in figure 2. A detailed description of the triaxial apparatus is presented in works [7, 8].

The result of the test is a graph of the recorded voltage on the piezoelectric bender elements, which is the basis for determining the wave velocity. On the basis of the P-wave propagation time, the value of the modulus of elasticity M can be calculated from the following equation:

$$M = \rho V_p^2 \tag{1}$$

where:
ρ – bulk density,
Vp – P-wave velocity.

The time of the wave propagation is measured by the "peak to peak" method. Cylindrical samples, 5 cm in diameter and 5 cm high, were used. Before starting the velocity measurement, delay time and polarity were checked. For a test sample, a series of tests was performed to determine the correct amplitude and frequency of the input signal for the shear and compression waves [9]. All tests were performed using a sinusoidal wave,

DOI: 10.1051/e3sconf/20172402006

generated with time intervals of 15 seconds. This time was determined on the basis of the total time of signal suppression and the time necessary for registering the reply. The registered signal was filtered into a continuous function to better evaluate wave transition time. For this purpose, a moving average method was used with an eight-point window. For each sample, twenty designations were made. In addition, for comparison purposes and the possibility of modelling the medium in subsequent research, measurements of the S-wave velocity were carried out to determine the dynamic modulus of rigidity G:

$$G = \rho V_s^2 \tag{2}$$

where: Vs – S-wave velocity.

Assuming the elastic model of the analysed soil and knowledge of shear and pressure wave velocity, it is also possible to determine the dynamic Poisson's ratio □, according to the formula.

$$\nu = \frac{\frac{1}{2}\left(\frac{V_P}{V_S}\right)^2 - 1}{\left(\frac{V_P}{V_S}\right)^2 - 1} [-] \tag{3}$$

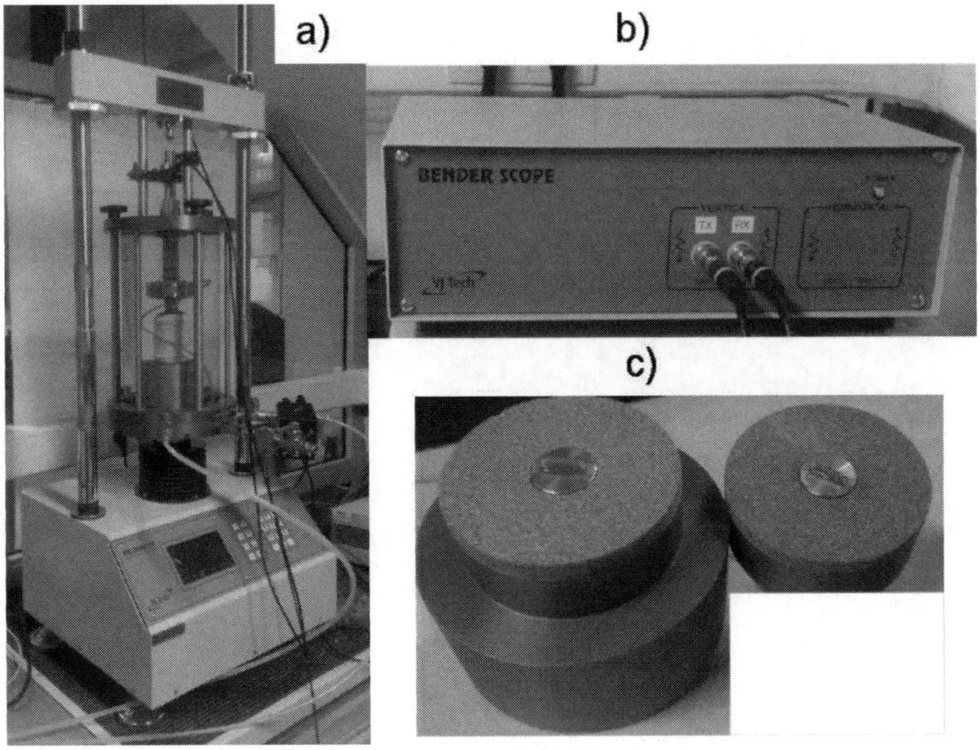

Fig. 2. Test equipment a) triaxial apparatus b) waveform generator c) bender element.

At the beginning, series of sample tests were conducted for identification the best signal in the receiver, which allows a precise determination of wave time [9].

The tests of the natural moisture samples were carried out for the S-wave at a 10V amplitude and a signal frequency in the transmitter of 1kHz and for P-wave the amplitude was 12V and frequency 5kHz. Furthermore, the tests of the natural, dried sample were carried out for the S-wave for 10V amplitude and the frequency of 10kHz, for P-wave the amplitude was 12V and frequency 5kHz.

The test time was adjusted in order to reach the peak and to mute the signal in the receiver. The signal was filtered to a continuous function, in order to obtain a more precise evaluation of the wave propagation time. Twenty tests were performed for each sample. The applied filtering method is a moving average method called the "running mean smoother," which uses an eight-point window.

4 Results and analysis

The measurement results for the natural moisture sample are shown in figures 3 and 4, and the measurement results for the dry sample are shown in figures 5 and 6. The measured and calculated parameters are shown in table 1 for P-wave and in table 2 for S-wave.

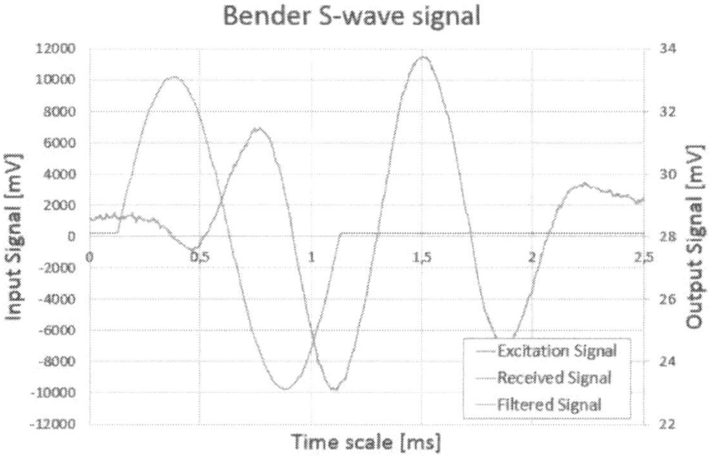

Fig. 3. The S-wave excitation and received signal of natural water content sample.

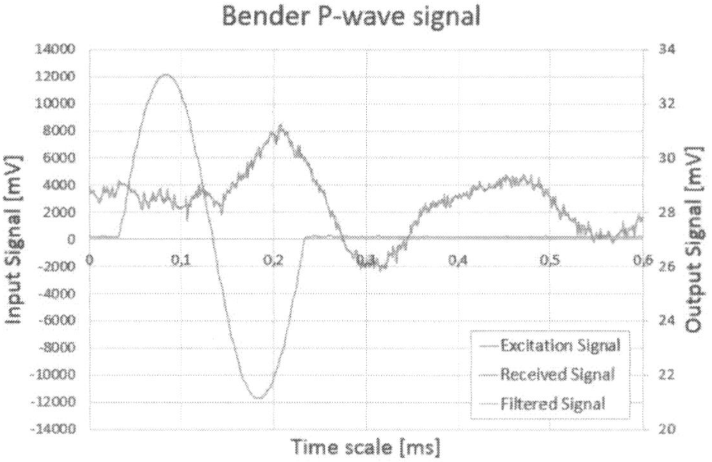

Fig. 4. The P-wave excitation and received signal of natural water content sample.

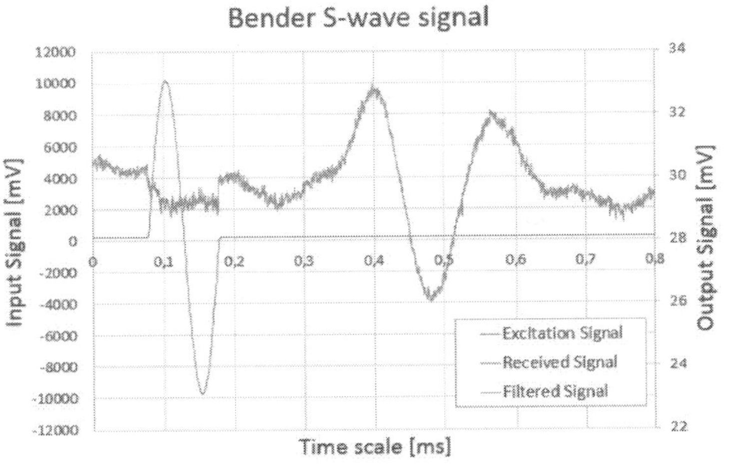

Fig. 5. The S-wave excitation and received signal of a dry sample.

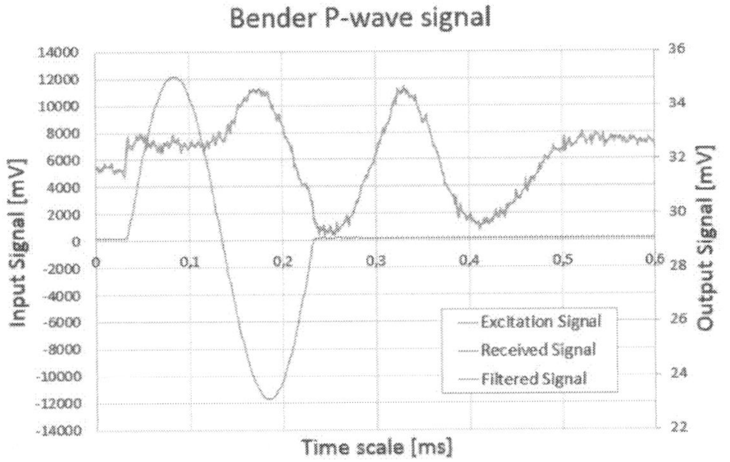

Fig. 6. The P-wave excitation and received signal of a dry sample.

Table 1. The average values of the parameters of the tested samples for P-wave test.

Sample	Bulk density [kg/m3]	P-wave velocity [m/s]	M modulus [MPa]
Natural water content	1780	221	87
Dry sample	1780	316	178

Table 2. The average values of the parameters of the tested samples for S-wave test.

Sample	Bulk density [kg/m3]	S-wave velocity [m/s]	G modulus [MPa]
Natural water content	1780	101	18
Dry sample	1780	82	12

For the determined P-wave, the difference in the calculated values for the dry sample and for the natural moisture sample is rather high. In the case of the S-wave, the water impact is lower. The dynamic G modulus of the soil samples after drying insignificantly decreased, which may be due to a change in the structure of the sample which, after drying, had a fine sand/dust structure.

In the case of the P-wave, a significant increase in the dynamic M modulus was observed after drying. It is possible that it is the result of the sample compaction after drying. Poisson's coefficient for soil samples with a natural water content was 0.37. For dry soil sample, the coefficient was 0.46.

5 Conclusions

This work presents the results of the laboratory tests of the dynamic modules of soil samples from a landfill site in the area of the closed Solvay Sodium Plant in Krakow. The results of laboratory tests, together with the results of field tests, were carried out as a part of the diagnosis of the ground for the investment of the John Paul II Centre in Łagiewniki, Krakow. They will be used to build a numerical model which will allow to analyse the impact of traffic load on the buildings of the John Paul II Centre.

The study area is characterized by a high variability of the subsoil state, and, in particular, its consistency. Due to the possibility of using the subsoil as a foundation for the future road investment, extensive tests on its stiffness should be conducted. This article discusses the lab methodology for determining the stiffness of the subsoil, expressed by the dynamic modulus of elasticity M and rigidity G. For this purpose, a triaxial apparatus equipped with piezoelements was used.

The preliminary results show relatively small values of the subsoil dynamic stiffness parameters. This points to the need to strengthen this subsoil for its use for construction purposes, and, in particular, for the construction of a future transportation route.

References

1. I. Krzak, *Zagospodarowanie terenów poprzemysłowych Krakowskich Zakładów Sodowych „Solvay"*, Problemy Ekologii Krajobrazu, Tom 17, 283-287 (2005)

2. B. Czado, *Analiza nośności pali fundamentowych na podstawie polowych badań gruntów sondą statyczną*, Rozprawa doktorska, Biblioteka Politechniki Krakowskiej (2014)

3. A. Ślęzak, *Wpływ składowiska odpadów Krakowskich Zakładów Sodowych na wody* [In]: *Przewodnik III Konferencji Sozologicznej PTG. Sozologia na obszarze antropopresji - przykład Kraków* (Wyd. AGH, Kraków, 1993)

4. Biuro Rozwoju Krakowa, *Miejscowy plan zagospodarowania przestrzennego terenów Krakowskich Zakładów Sodowych „Solvay" w likwidacji, w rejonie ulic: Zakopiańskiej, Myślenickiej i Podmokłej* (Kraków, 1994)

5. *Program Funkcjonalno-Użytkowy „Budowa Trasy Łagiewnickiej w Krakowie od skrzyżowania z ul. Grota-Roweckiego do skrzyżowania z ul. Beskidzką i z ul. Halszki wraz z budową odcinka linii tramwajowej"*. Materiały przetargowe, Spółka Trasa Łagiewnicka S.A

6. T. Godlewski, T. Szczepański, *Metody określania sztywności gruntów w badaniach geotechnicznych* (Wydawnictwo Instytut Techniki Budowlanej, Warszawa, 2015)

7. J. Zięba, D. Szwarkowski, *Slope stability analysis of an open-pit mine as a prospective S-7 and S-52 road intersection*, Czasopismo Techniczne (2017) (in press)

8. J. Kogut, J. Zięba, *The measurement of selected soil parameters of former open pit mine with the use of triaxial stress apparatus*, Czasopismo Inżynierii Lądowej, Środowiska i Architektury **1**, 64, 97-104 (2017)

9. W. Świdziński, J. Mierczyński, *Investigations of elastic response of non-cohesive soils by seismic waves*, Czasopismo Techniczne. Środowisko **107**, 1-Ś, 63-82 (2010)

10. http://trasalagiewnicka.krakow.pl/

11. http://www.geoportal.gov.pl//

E3S Web of Conferences **24**, 02007 (2017) DOI: 10.1051/e3sconf/20172402007
AG 2017 - 3rd International Conference on Applied Geophysics

Impact of rock mass temperature on potential power and electricity generation in the ORC installation

Michał Kaczmarczyk[1*]

[1]AGH University of Science and Technology, 30-059 Krakow, Poland

Abstract. The basic source of information for determining the temperature distribution in the rock mass and thus the potential for thermal energy contained in geothermal water conversion to electricity, are: temperature measurements in stable geothermic conditions, temperature measurements in unstable conditions, measurements of maximum temperatures at the bottom of the well. Incorrect temperature estimation can lead to errors during thermodynamic parameters calculation and consequently economic viability of the project. The analysis was performed for the geothermal water temperature range of 86–100°C, for dry working fluid R245fa. As a result of the calculations, the data indicate an increase in geothermal power as the geothermal water temperature increases. At 86°C, the potential power is 817.48 kW, increases to 912.20 kW at 88°C and consequently to 1 493.34 kW at 100°C. These results are not surprising, but show a scale of error in assessing the potential that can result improper interpretation of the rock mass and geothermal waters temperature.

1 Introduction

Research of deep geothermal systems requires careful analysis of a number of key geological factors, where the quality of the data determining the reliability of the models [1-3]. In the case of thermodynamic analysis of heat accumulated in geothermal waters utilization [4-5], ex. using Organic Rankine Cycle (ORC), one of the key parameters is the temperature of geothermal waters determined by the thermal parameters of the rock mass. This is crucial in the context of correct estimation of power, efficiency and quantity of electricity. Even a slight error of several Celsius degrees, can result in the underestimation or overestimation of potential that can be possible use and leads into an incorrect economic assessment of the planned installation. Therefore, it is important to interpret accurately the thermal data of the rock mass. The paper presents the consequences of incorrect determination of the temperature of geothermal waters used in the ORC systems.

* Corresponding author: mkz@agh.edu.pl

2 Temperature data

The primary source of information in the case of in-depth temperature -distribution, and hence the potential for conversion of heat accumulated in geothermal waters into electricity are temperature measurements in stable geothermic conditions, temperature measurements in unstable conditions, and maximum temperature measurements at the bottom of the borehole [6].

From the point of view of correct determination of geothermal waters temperature, the measurement in stable conditions (thermal equilibrium) are crucial. In the case of measurement in unstable conditions, we have to deal with disrupted temperature distribution which results from the drilling process, mainly from the long-term drilling fluid circulation [6-7]. It causes growth of the temperature in the shallower zone of the borehole and its lowering at the bottom of the borehole, which in consequence leads to incorrect interpretation of the temperature log. According to Plewa [8], it is possible to assume that the minimum time to reach thermal equilibrium after drilling is finished is 7 days for boreholes with depth up to 1 500 m, and 12–14 days for boreholes up to 3 000 m deep. Because the circulation of the drilling fluid is the shortest in the deepest borehole interval, the data from the maximum temperature measurements on the bottom of the borehole can also be useful [6]. The above also leads to the conclusion that in the thermal profile there is an interval in which the temperature corresponds to the real values (Figure 1), but it is impossible to determine that zone [6].

A different issue is the impact of paleoclimatic changes on which detailed information can be found in publications [7, 9-11].

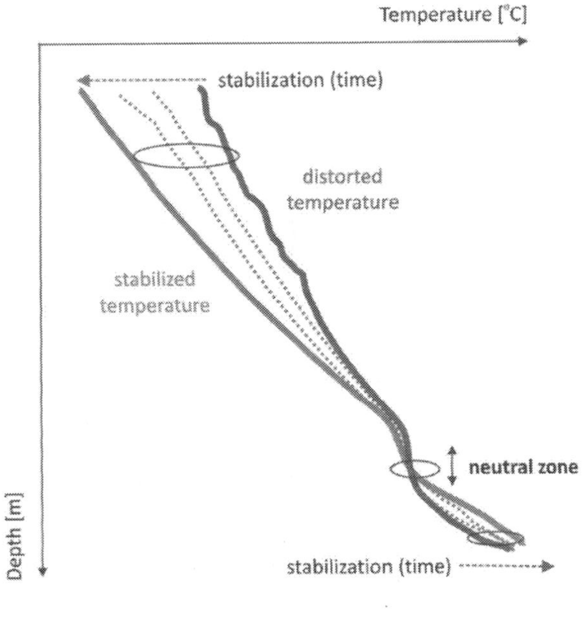

Fig. 1. Temperature changes caused by disturbances of thermal equilibrium in wells (based on [6]).

The use of temperature data measure in unstable conditions can lead to errors in estimating the power and efficiency of geothermal power plants, and consequently to the errors in the key economic cost of the project to be implemented from the investor's point of view.

3 Estimation of potential power and conversion efficiency of heat into electricity in ORC geothermal power plants

3.1 Technology of conversion heat into electricity

The technology of converting electricity using geothermal energy depends on the thermal and hydrogeological conditions prevailing in the area. This specificity has led to the emergence of a variety of technologies that allow the generation of electricity based on geothermal resources.

The use of organic working fluid is the most common method of producing electricity using geothermal waters with the temperature not exciding 100°C [12] and hot dry rocks reservoir [13]. Properly selected working fluid has boiling temperature lower than geothermal water temperature, at a pressure higher than the ambient pressure. It works according to the Clausius-Rankine comparative cycle [14].

A typical work cycle performed by geothermal water and the working fluid at the ORC power plant is shown on Figure 2. The geothermal water extracted through the production well is directed by particle filter to the heat exchanger system: superheater-evaporator-preheater, which transfers heat accumulated in geothermal water into the working fluid and evaporate it. After that the working fluid is moved as a steam to the turbogenerator, whereby by changing the potential energy accumulated in the working fluid into mechanical energy of the turbine blades rotation, electricity is produced. The next step is directed the working fluid vapor to the condenser and liquefy it to the original state (liquid), which is the closure of the cycle. Condensation is possible by directing to the condenser cooling water, which receiving heat from the working fluid. For this process to be carried out, an open cooling circuit such as surface water, or alternatively a cooling tower, is used.

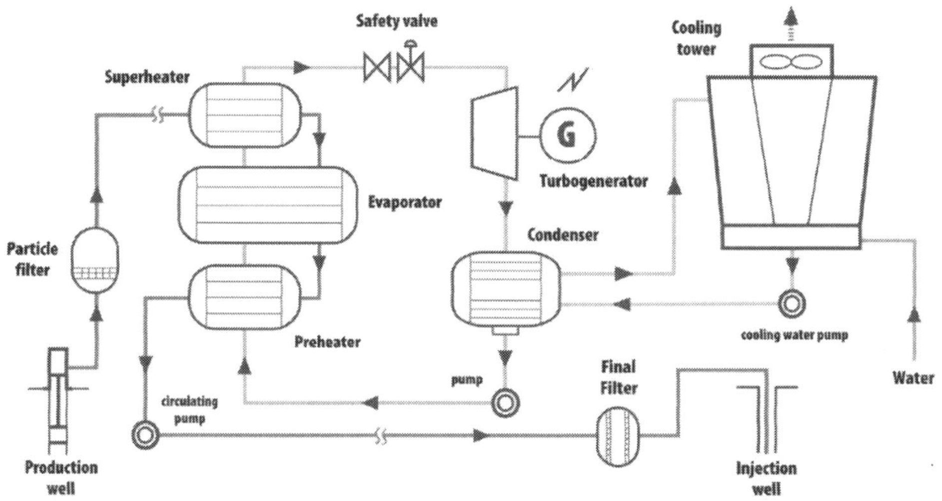

Fig. 2. A schematic diagram of an ORC plant (based on [12]).

3.2 Geothermal water as a determinant of efficiency and power

The main factors influencing the thermal efficiency and power of a geothermal power plant are: **geothermal water temperature** at outflow, geothermal water flow rate, physicochemical properties of geothermal water, type of technology used, condensation temperature and cooling system used. At the same time, the temperature of the geothermal water supplying the geothermal power plant is crucial for both its power and efficiency, which leads to the conclusion that these parameters increase with increasing geothermal water temperature. Rock mass temperature, and hence geothermal water temperature, are determined by the density of the Earth's heat flux and by the thermal properties of the rocks (thermal conductivity, specific heat) in the studied lithological profile. It can be determined primarily on the basis of a detailed analysis of drilling data, which the geological structures have been drilled and identified.

4 Assumptions and methodology of performing thermodynamic calculations

The analysis was performed for the geothermal water temperature range from 86°C to 100°C, changing the temperature range every 2°C. This range is due to the temperature of geothermal waters available for use in Poland. In the presented studies on the potential of geothermal systems based on ORC, dry working fluid R245fa was analyzed, which according to previous analyzes by the author [15], allows to obtain the highest efficiency and power of the ORC geothermal power plant. For the calculations author developed the thermodynamic model, with the values of temperature, pressure, enthalpy and entropy at specific points determined using the NIST REFPROP 9.1 database. This is a program developed by the National Institute of Standards and Technology for the analysis of thermodynamic processes [15].

For the calculation model, it was assumed that the Organic Rankine Cycle consists of a turbine (connected to a generator), a condenser (connected to a cooling tower), a feed pump, a preheater, an evaporator and a superheater (Figure 3). The power of a geothermal power plant can be described in accordance with the first principle of thermodynamics [12, 17]:

$$W_{gross} = Q_d \bullet \eta_{ORC} = m_{wf} \bullet (h_{wf1} - h_{wf4}) \bullet \eta_{ORC} \tag{1}$$

where:
W_{gross} – power [W]
Q_d – heat flux supplied to the superheater, evaporator and preheater [W]
η_{ORC} – thermal efficiency [-]
m_{wf} – mass flow of the working fluid[kg/s]
h_{wf1}, h_{wf4} – specific enthalpy [kJ/kg]

The amount of gross electricity was calculated from the power plant capacity and the working time. Based on the analysis of the data contained in the "Atlas ..." [18], it was assumed that the plant's operating time would be 4 000 hours:

$$E_{el} = W_{gross} \bullet t \tag{2}$$

where:
E_{el} – electricity [Wh]
W_{gros} – power [W]
t – operation time [h]

The obtained results refer to the gross power and the gross electricity output. The methodology of calculations was adopted from [12, 16, 18].

For calculations the values corresponding to geothermal water from the borehole of the Bańska PGP-1 were assumed. The flow rate of the production well was taken as the maximum value 550 m^3/h. It was assumed that the temperature of the geothermal water leaving the geothermal power plant would be 60°C. The working fluid condensation temperature was assumed as a constant value of 30°C, as well as the overheating temperature of the operating medium T_s=3°C. The mineralization of geothermal water was assumed to be a constant value of 3.12 g/dm^3. Due to the influence of geothermal water temperature parameter, the values of the geothermal water density have been estimated based on the methodology given in [19-20] in a variable temperature range of 86–100°C. This is one of factors which determine the conditions in which geothermal waters are used, especially the amount of energy obtained and the manner in which cooled water is utilized [5].

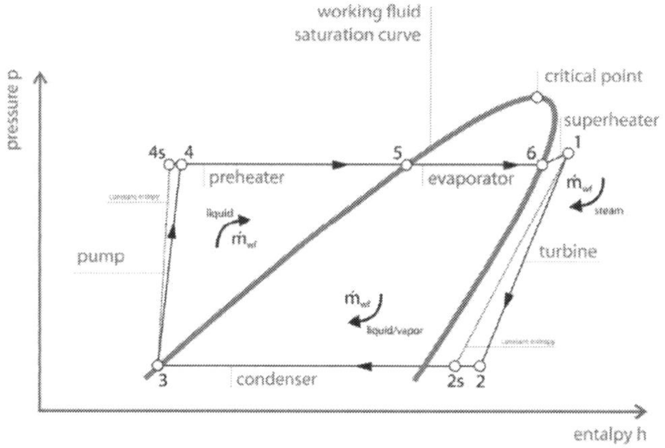

Fig. 3. Pressure enthalpy diagram with applied specific points; Explanation: 1–2s – isentropic expansion of the working fluid in the turbine, 2–2s – cooling of the working fluid in the condenser, 2s–3 – condensation of the working fluid in the condenser, 3–4s – isentropic pumping of the working fluid by the feed pump, 5 – heating of the working fluid in the preheater, 5–6 –evaporating of the working fluid in the evaporator, 6–1 – superheating of the working medium in the superheater (based on [12]).

5 Results and analysis

The calculations show that the geothermal water temperature is crucial for the thermal efficiency and power output. From the point of view of proper assessment of the rock parameters, a correct assessment of thermal parameters should be a priority, as having a direct impact on the estimation of the potential use of energy accumulated in geothermal waters for electricity purposes. The data compiled in Table 1 and on Figures 3 and 4 show the significant increase in geothermal power as the geothermal water temperature increases. At 86°C, the potential gross power is 817.48 kW, increases to 912.20 kW at 88°C and consequently to 1 493.34 kW at 100°C. These results are not surprising, but show a scale of possible error in assessing the potential that can result improper interpretation of the rock mass temperature. Similar as with power, efficiency also increases from 6% (at 86°C) to 8% at 100°C. Power and efficiency determine the amount of electricity that increases from 3 269.92 MWh to 5 973.34 MWh in the studied temperature range 86–100°C.

The data show that in case of an error in estimating the temperature of geothermal waters we are dealing with overestimation or underestimation of potential capacity and electricity of geothermal power plant by c.a. 11% for every +/- 2°C. It should be emphasized that this type of error results in the selection of individual components of the plant, such as the size of the turbine, heat exchangers or the power of the feed pump, and thus the optimization of the system so that it operates at optimum performance using the full potential of the geothermal resource.

Table 1. Results of thermodynamic calculations.

Temperature	Geothermal water density	Power	Electricity	Efficeincy	Turbine / feed pump power
°C	kg/m^3	kW	MWh	-	kW
86	990,8162	817,48	3 269,92	0,06	1 044,01 / 23,23
88	989,5068	912,20	3 648,78	0,07	1 183,06 / 26,88
90	988,1806	1 007,78	4 031,10	0,07	1 327,48 / 30,77
92	986,838	1 103,57	4 414,29	0,07	1 477,58 / 36.06
94	985,4792	1 200,02	4 800,07	0,07	1 632,11 / 40,47
96	984,1043	1 298,15	5 192,62	0,08	1 793,73 / 45,10
98	982,7136	1 396,34	5 585,36	0,08	1 961,59 / 51,29
100	981,3072	1 493,34	5 973,34	0,08	2 133,04 / 57,82

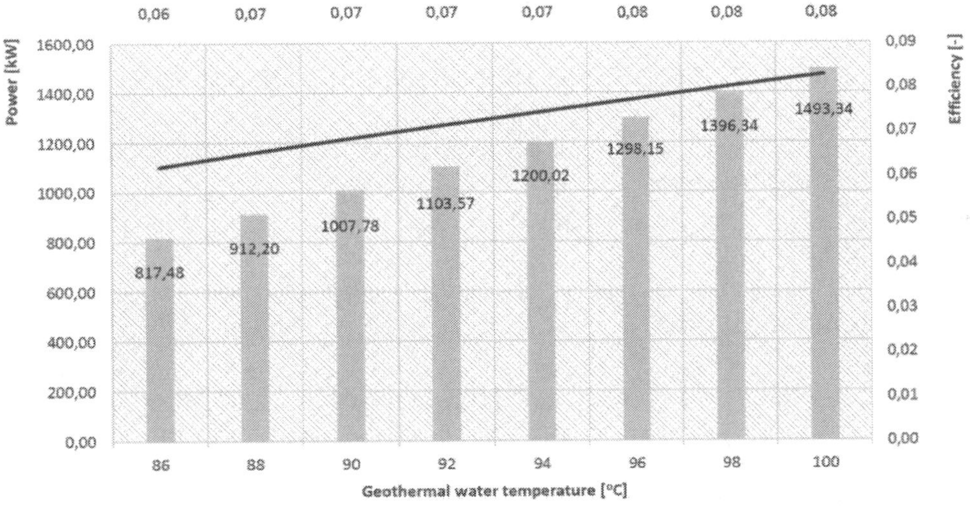

Fig. 3. The results of the thermodynamic calculations for the obtained power and efficiency.

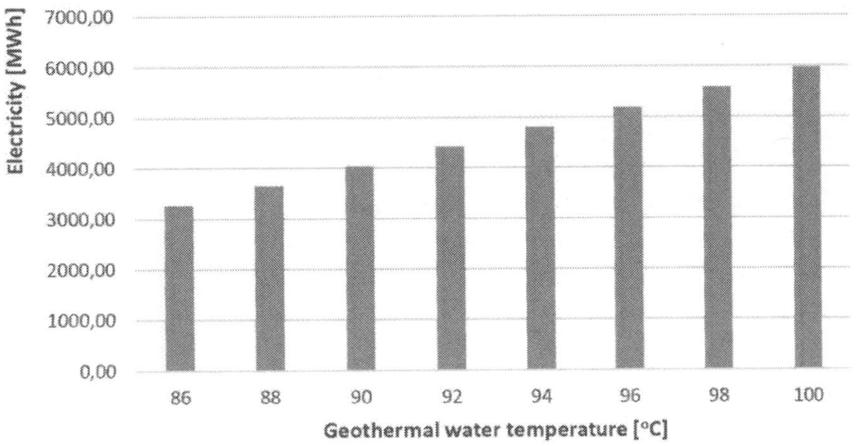

Fig. 4. The amount of gross electricity depends on the geothermal water temperature.

6 Conclusions

1. Research of deep geothermal systems requires careful analysis of a number of key geological factors, especially temperature which has great impact on the potential power, efficiency and amount of electricity in ORC power plant.
2. From the point of view of correct determination of geothermal waters temperature, the measurement in stable conditions (state of geothermic equilibrium) are crucial. The use of temperature data measure in unstable conditions will lead to errors in estimating the power and efficiency of geothermal power plants, and consequently to the errors in the key economic cost of the implemented project from the investor's point of view.
3. Making an error in estimating the temperature of geothermal waters we are dealing with overestimation or underestimation of of potential capacity and electricity of geothermal power plant by c.a. 11% for every +/- 2°C.

References

1 W. Bujakowski, B. Tomaszewska, M. Miecznik, Renew. Energy **99,** 420–430 (2016) doi 0.1016/j.renene.2016.07.028

2 M. Pussak, K. Bauer, M. Stiller, W. Bujakowski, J. Appl. Geophys. **103,** 186-198 (2015) doi 10.1016/j.jappgeo.2014.01.020

3 B. Tomaszewska, L. Pająk, G. Hołojuch, Desalin. Water Treat. **69**, 316–321 (2017) doi: 10.5004/dwt.2017.0657

4 B. Tomaszewska, L. Pająk, Gospod. Surowcami Min. **4**, 59–70 (2012) doi: 10.2478/v10269-012-0038-7

5 B. Tomaszewska, L. Pająk, *Using treated geothermal water to replenish network water losses in a district heating system*, Pol. J. Environ Stud., **22,** 1, 243-250 (2013)

6 J. Szewczyk, M. Hajto, *Strumień cieplny a temperatury wgłębne na obszarze Niżu Polskiego* [in.] W. Górecki [ed.], *Atlas zasobów geotermalnych formacji mezozoicznej na Niżu Polskim* (Kraków 2006)

7 J. Szewczyk, *Estymacja gęstości strumienia cieplnego metodą modelowań właściwości termicznych ośrodka*, Przegląd Geologiczny **49**, 11 (2001)

8 S. Plewa, *Rozkład parametrów geotermalnych na obszarze Polski* (Wydawnictwo CPPGSMiE PAN, Kraków 1994)

9 J. Majorowicz, J. Safanda, R. Przybylak, G. Wójcik, *Rekonstrukcja zmian temperatury powierzchni gruntu w Polsce w ostatnim 500-leciu na podstawie profili geotermicznych*, Przegląd Geologiczny **4** (2001)

10 J. Majorowicz, M. Wróblewska, P. Krzywiec, *Interpretacja i modelowanie ziemskiego strumienia cieplnego w obszarze eksperymentu sejsmicznego POLONAISE'97 – analiza krytyczna*, Przegląd Geologiczny **50**, 11 (2002)

11 J. Szewczyk, *Ślady zmian klimatycznych plejstocenu oraz holocenu w profilach temperatury w głębokich otworach wiertniczych na Niżu Polskim*, Przegląd Geologiczny **50**, 11 (2002)

12 R. DiPippo, *Geothermal Power Plants: principles, applications, case studies and environmental impact* (Library of Congress Cataloging-in-Publication Data, Elsevier Ltd. 2012)

13 W. Bujakowski, M. Miecznik, L. Pająk, R. Skrzypczak, A. Sowiżdżał, Renew. Energy **80**, 441–453 (2015) doi.org/10.1016/j.renene.2015.02.018

14 A. Borsukiewicz-Gozdur, W. Nowak, A.A. Stachel, *Systemy do generacji prądu elektrycznego przy wykorzystaniu wód geotermalnych – perspektywy rozwoju w Polsce*, Materiały Ogólnopolskiego Kongresu Geotermalnego, Radziejowice (2007)

15 M. Kaczmarczyk, *Analiza i ocena możliwości produkcji energii elektrycznej przy wykorzystaniu energii cieplnej zakumulowanej w wodach geotermalnych na obszarze województwa małopolskiego*, rozprawa doktorska (AGH WGGiOŚ KSE, Kraków 2016)

16 A. Borsukiewicz-Gozdur, W. Nowak, *Desirable Thermophysical Properties of Working Fluids In Organic Rankine Cycle*, Proceedings European Geothermal Congress (Unterhaching, Germany 2007)

17 E.W. Lemmon, M.L. Huber, M.O. McLinden, *NIST Reference Fluid Thermodynamic and Transport Properties – REFPROP, Version 9.1, User's Guide*, U.S. Department of Commerce, Technology Administration, (National Institute of Standards and Technology, Standard Reference Data Program 2013)

18 W. Bujakowski W., B. Tomaszewska [eds.], *Atlas wykorzystania wód termalnych do skojarzonej produkcji energii elektrycznej i cieplnej przy zastosowaniu układów binarnych w Polsce* (IGSMiE PAN, Kraków 2014)

19 M. H. Sharqawy, V. J. H. Lienhard, S. M. Zubair, Desalin. Water Treat. **16**, 354–380 (2010) doi 10.5004/dwt.2010.1079

20 M. Miecznik, *Błąd szacowania potencjału dla wytwarzania energii elektrycznej w instalacjach binarnych typu ORC związany ze zmiennością parametrów termodynamicznych wody geotermalnej*, Technika Poszukiwać Geologicznych Geotermia, Zrównoważony Rozwój **2**, 155-166 (2013)

Comparison of an empirical S-wave velocity model and a calculated stress-strain model for a rock mass disturbed by mining

Krzysztof Krawiec[1*], *Rafał* Czarny[1]

[1]Mineral and Energy Economy Research Institute, Polish Academy of Sciences, 31-261 Krakow, Poland

Abstract. In the article a comparison analysis is presented between a numerical model of the stress and deformation state in a rock mass and an S-wave velocity model obtained as a result of in situ measurement. The research was conducted using data from the Jastrzębie and Moszczenica coal mines. The part of the rock mass examined was strongly disturbed by multi-seam exploitation of coal. To obtain the S-wave velocity model 6 hours of ambient seismic noise data were recorded using 11 seismometers. The propagation of the Rayleigh surface wave between the seismometers was reconstructed utilising the seismic interferometry and the cross correlation technique. Estimation of a two dimensional model of the S-wave velocity field was performed on the basis of dispersion curves of the Rayleigh wave phase velocity. The stress and deformation field were calculated assuming a plane state of stress with the use of the elastic-plastic Coulomb-Mohr strength criterion. Images of the vertical stress, horizontal stress, vertical strain and horizontal strain as well as the subsidence profile on the model surface were obtained as a result of the calculation. Analysis of the results shows correlation between the field of S-wave velocity and the modelled field of stress and strain.

1 Introduction

Information about the stress and deformation processes developing in a rock mass disturbed by underground mining is essential in terms of the terrain surface subsidence and its effect on structures and the environment [1-3]. Moreover, strong seismicity may be induced in zones with a high stress gradient [4, 5].

One of the parameters that describe the stress and deformation changes in the rock mass is the velocity of seismic waves [6]. In mines this parameter has mainly been determined by seismic profiling [7, 8] or active and passive seismic tomography [9-11]. Recently, seismic interferometry has also been used for imaging [12, 13] and monitoring of the rock mass influenced by mining [14, 15]. This cost-effective tool can turn seismic noise into a useful signal produced by a so-called "virtual seismic source" [16].

* Corresponding author: k.krawiec@meeri.pl

 DOI: 10.1051/e3sconf/20172403001

A wide range of numerical modelling codes has been used to investigate the stress and deformation changes induced by underground coal mining [17]. Among these the FLAC code is a well-known computer code for such analysis of rock engineering problems using the FVM/FDM approach with a vertex scheme of triangular and/or quadrilateral grids[18, 19]. Application of FLAC codes in the modelling of the exploitation of multiple coal seams was widely described by Kwaśniewski and Wang [20], Kook et al. [21], Han et al. [22], Pilecki [23], Wesołowski [24], Zhang et al. [25], Sui et al. [26], Zhu et al. [27].

In this paper, the measured S-wave velocity field and numerically modelled stress-deformation field of a rock mass disturbed by the exploitation of multiple coal seams are compared. In the initial part of the paper, we show the location of the research area and we present brief information about the geological and mining conditions in the Jastrzębie and Moszczenica coal mines in Poland. Then, the methodology of the S-wave velocity model using seismic interferometry with ambient seismic noise is explained. Next, we describe the methodology of numerical calculations of stress and deformation fields.

Finally, we analyse the characteristic features of the empirical seismic velocity model comparing it to the calculated stress and deformation model. The conclusions formulated correspond to the relationship between changes in S-wave velocity and the stress and deformation processes induced by mining.

2 Geological and mining conditions

The research area is located at Jastrzębie and Moszczenica mines near to the town of Jastrzębie-Zdrój in southern Poland (Fig. 1a). The locations of profiles M1-M1' for numerical calculations and profile P1-P1' for the empirical S-wave velocity model, together with information about the geological and mining conditions, are shown in Figure 1b. Profiles P1-P1' and M1-M1' were located across a pillar used as a protective boundary between coal mines (Fig. 1b, red dashed line). Underground mining of coal has taken place in the research area since the 1960's (Fig. 1b, white line).

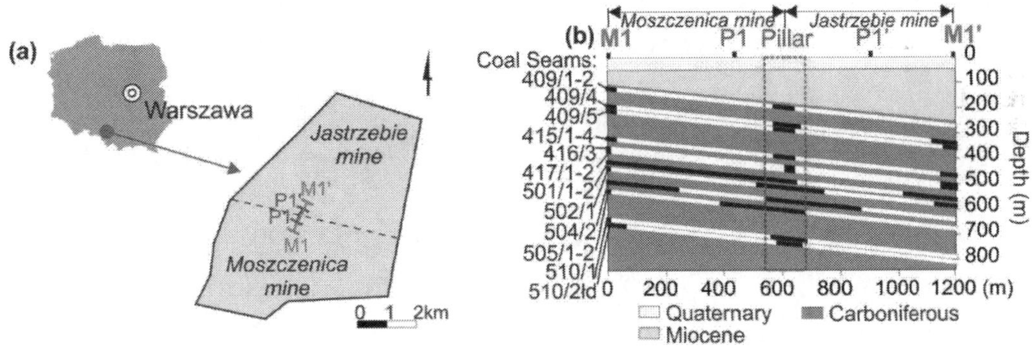

Fig. 1. Location of numerical modelling profile M1-M1' and S-wave velocity profile P1-P1' at Jastrzębie and Moszczenica mines; (b) Geological cross-section for M1-M1' and P1-P1' profiles. The white and black lines indicate exploited and unexploited coal seams, respectively. The red dashed line indicates the pillar between the Jastrzębie and Moszczenica mines (based on [30]).

The Carboniferous formation is composed of clastic rocks such as sandstones, mudstones and hard coal seams. This formation dips from south to north at an average inclination of 7 degrees. Miocene and Quaternary sediments occur above these seams from a depth of about 170 m to the ground surface. These sediments are composed of gravels, sands, clay and sandstones.

3 Methodology

3.1 S-wave velocity model

An S-wave velocity model was calculated on the basis of seismic interferometry with the use of ambient seismic noise. The array contained 11 broadband seismometer stations Guralp CMG-6td. The stations were deployed on the ground surface along profile P1-P1' (Fig. 2a). The time of 6 hours of ambient seismic noise data with 100 Hz frequency sampling were recorded and processed according to the methodology presented by Bensen et al. [28]. The main processing procedures contained: detrend, bandpass filtering (0.5–10 Hz), 1-bit normalisation, spectral whitening and cross-correlation of all available combinations of station pairs. The example of cross-correlation functions (CCFs) for station G1 as a virtual seismic source is shown in figure 2b.

Because the main energy of seismic noise comes from the northerly direction from traffic on active roads, working mine surface machines etc., the only positive part of the CCFs shows a Rayleigh surface wave (Fig. 2b). Based on various combinations of CCFs, 14 dispersion curves were identified along profile P1-P1' (Fig. 2c, d). Each dispersion curve was inverted to a 1D S-wave velocity model by using the gradient algorithm of Xia et al. [29]. The depths of the main geological formation (Fig 1b) were used in the inversion procedure. Finally, the 2D S-wave velocity model along profile P1-P1' was interpolated and correlated with the geology and mining information (Fig. 3).

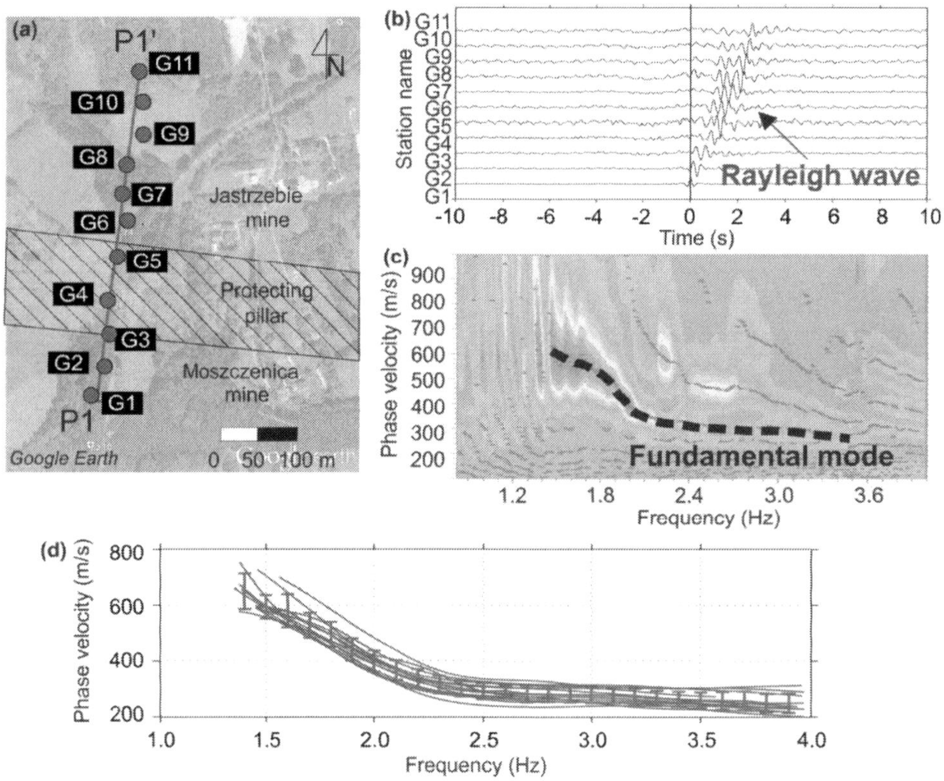

Fig. 2. (a) Location of stations G1-G11 along the profile P1-P1'; (b) an example of CCFs for a virtual source in station G1; (c) an example of the identification of the Rayleigh wave dispersion curve; (d) all dispersion curves computed along profile P1-P1' (based on [30]).

3.2 Numerical modelling

3.2.1 Physical model

The physical model was constructed on the basis of the geological cross-section along the M1-M1' profile (Fig. 1b). The model was 1200 m long, 840 m deep and consisted of 29 geological layers. The 12 coal seams were included in the model with the exploited working faces marked.

The information about the material properties was established on the basis of analogy with data from the coal mine and expert opinion. Averaged values of the material properties for each layer were assumed when carrying out the calculations (Tab. 1).

Table 1. Material properties.

	Miocene	Sandstone	Shale	Hard coal
Young's Modulus [GPa]	1.5	19.3	11.1	0.75
Poisson ratio [-]	0.25	0.38	0.29	0.25
Bulk density [kg/m^3]	2000	2700	2700	1922
Cohesion [MPa]	0.3	27.2	38.4	0.15
Tension [MPa]	0	1.17	14.4	0
Angle of internal friction [deg]	38	27.8	14.4	20

3.2.2 Calculation model

Numerical calculations were conducted assuming a plane state of stress with the use of the elastic-plastic Coulomb-Mohr strength criterion. A two dimensional state of stress was modelled in a quasi-spatial nature. The two horizontal components of the main stress have equal values. The values of the primary vertical and horizontal stress components were changeable in the range of model depth [23]. On the laterals and the bottom edges of the model, displacement boundary conditions were established. The horizontal displacement was constrained along the vertical boundaries of the grid and no displacements were allowed at the base of the model. Motion of the nodes on the model surface was permitted in both directions. The dimensions of the grid elements approached the size of 2 x 2 m.

The calculations consist of 24 stages. In the first stage the model was calculated to obtain the equilibrium state assuming the gravitational distribution of primary stress. In subsequent stages, parts of the coal seams were excavated chronologically. After the excavation of one part of the coal seam, calculations were allowed until the vertical displacements in the roof reached the value of 95% of the height of the seam. The values of the properties of backfill material were equal to 75% of the values assumed for hard coal layers. In the next part, the model was calculated to obtain the equilibrium state of secondary stress. Images of vertical stress, horizontal stress, vertical strain and horizontal strain as well as the subsidence profile on the ground surface were obtained as a result of this calculation. The calibration of the numerical model was undertaken using a comparison of the subsidence curve along the M1-M1' profile with geodetic measurement data. The subsidence profile thus calculated approximately coincided with geodetic measurement data (Fig. 3). In the middle part of the curves one can see the location of the pillar on a profile with much smaller values of subsidence.

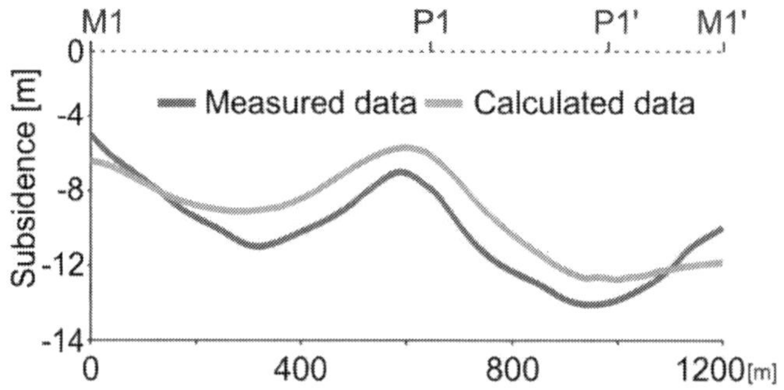

Fig. 3. The comparison of the subsidence curve for measured and calculated data.

4 Results and analysis

The S-wave velocity cross-section along the P1-P1' profile is presented in Figure 4. The largest velocity increase is observed in the top of the Carboniferous formation. Along the pillar (Fig. 4, between red dashed lines) and within the Quaternary and Miocene formations, the velocity decreases by about 100 m/s in comparison with the exploited area.

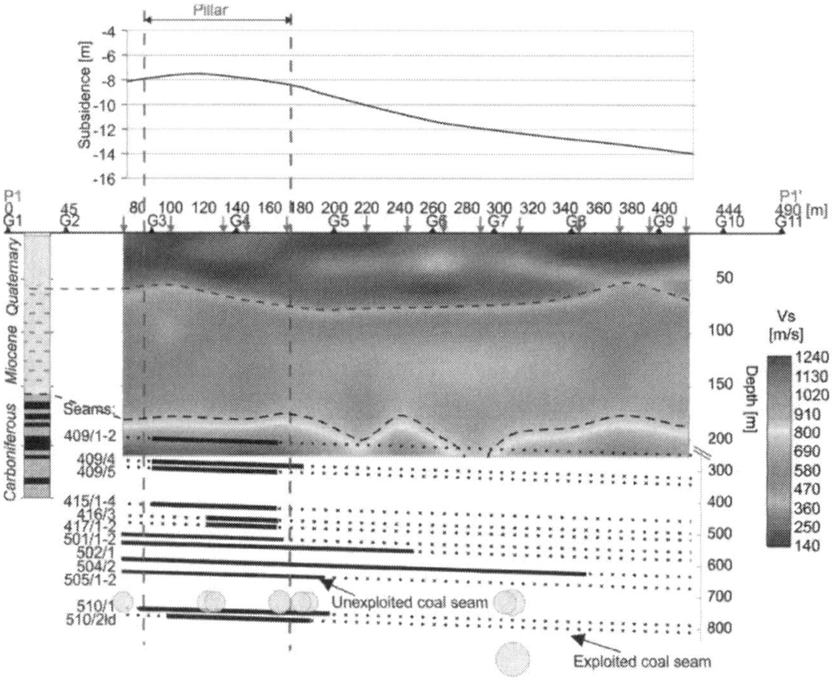

Fig. 4. An S-wave velocity model along the P1-P1' profile; the red arrows above the cross-section indicate locations of 1D S-wave profiles; the yellow dots indicate seismic events greater than 10^5 J occurring during the most recent mining activity in seams 510/1 and 510/2 (based on [30]).

This is caused by the deformation process of the rock mass due to exploitation. At the top of the Carboniferous layer the largest velocity with an average value of about 1200 m/s is seen in the pillar. In the same formation, but in the exploited area, the velocity shows a significant decrease with an average value of about 970 m/s. Generally the S-wave velocity field in the exploited area is much more complicated than within the pillar. The velocity increase in the pillar at the top of the Carboniferous layer can be caused by the concentration of stress and/or local geology. Moreover, a strong stress gradient around the pillar can induce seismic events. The irregularly-shaped border of the Carboniferous layer above the exploited area is the reason for mining multiple-seams of coal. The curve of subsidence correlates with the velocity of the Carboniferous formation.

The vertical stress field after the exploitation of the last coal seam 510/2łd is presented in Figure 5a. The vertical stress reaches the highest values of the order of -90 MPa (minus denotes compression) in the area of the safety pillar with coal remnants localised at a depth of about 440 m. The values are nine times greater than the primary stress values in that area. Concentration of compressive stress appears in almost the entire pillar and ranges from the floor of the Miocene formations to the bottom of the model. Beyond the pillar strong concentrations of compressive vertical stress are observed near the edge of the mined areas. The destressed zone appears in almost the entire section of the modelled rock mass in the vicinity of the exploited areas. The values of vertical stress are several times smaller than the values of the primary vertical stress. Destressed areas range from the surface to the bottom of the model at a depth of 840 m.

A comparison of the vertical stress field with the S-wave velocity field is presented in Figure 5c. The S-wave velocity increases in the area of higher values of compressive vertical stress in the pillar, which lies at a depth from 160 to 200 m.

The horizontal stress field, which is mostly compressive, is presented in Figure 5b. Strong concentration of horizontal stress appears mainly in sandstone layers and has much larger values than the primary horizontal stress. The maximum value of stress is about -186 MPa and appears in the roof layer of the 415/1-4 seam in the area of the pillar at a depth of 350 m. Zones of tensile horizontal stress which values reaches about 9 MPa are also observed in the model. These zones are located in the floor layers of the 409/5, 417/1-2 and 505/1-2 seams within the pillar. In addition, directly above the roofs of the exploited seams one can see a significant decrease in compressive stress values which sometimes change to tensile stress.

A comparison of the horizontal stress field with the S-wave velocity field is presented in Figure 5d. The highest values of compressive horizontal stress are concentrated within the pillar at a depth from 160 to 200 m. In the pillar area, where the values of stress are greater, the S-wave velocity increases.

The vertical strain field is presented in Figure 6a. The sign before the values of strain denotes the direction.The areas of highest values of vertical strain are localised in the zones of goaf. In addition, the fields of high values of strain appear within the pillar at a depth of from 60 to 180 m.

A comparison of the vertical strain field with the S-wave velocity field is presented in Figure 6c. The area of high values of strain within the pillar at a depth from 60 to 160 m is partially correlated with decreases in the S-wave velocity.

The horizontal strain field is presented in Figure 6b. The areas of the highest values of horizontal strain are localised within the pillar from surface of the model to a depth of 220 m below the ground surface. In addition, the fields of high values of strain appear near the sidewalls.

A comparison of the horizontal strain field with S-wave velocity field is presented in Figure 6d. The strain decrease at a depth of from 60 to 160 m within the pillar corresponds to a lesser increase in the S-wave velocity.

Based on the obtained results it can be summarised that in the areas with strong concentrations of stress and strain the S-wave velocity fluctuates. It could be connected with the changes in elastic properties of the rock mass.

DOI: 10.1051/e3sconf/20172403001

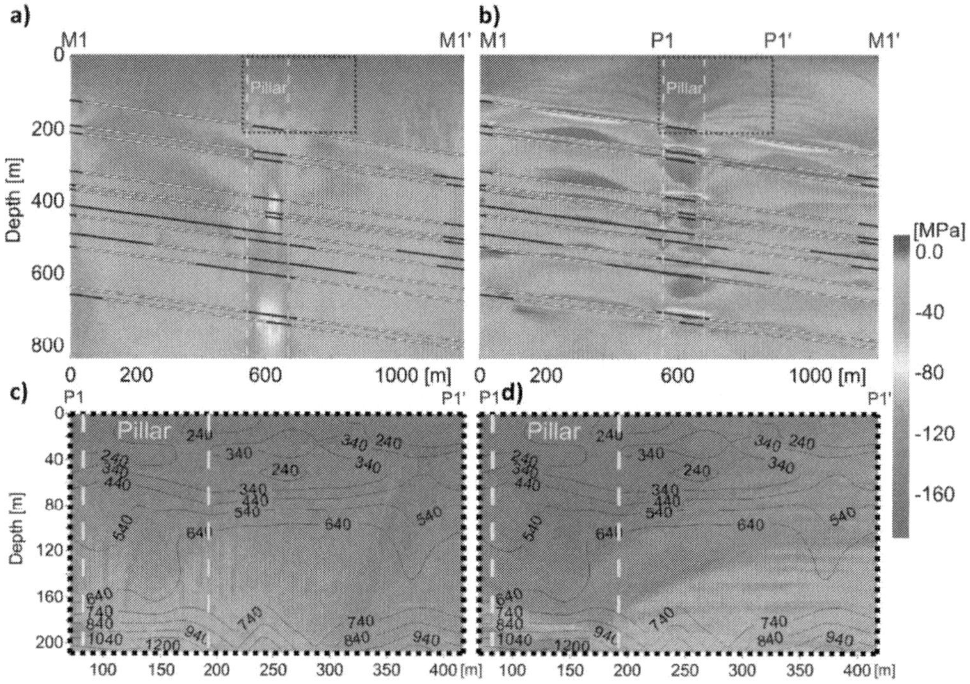

Fig. 5. (a) Vertical stress field, (b) Horizontal stress field (black dashed line indicates the range of the S-wave velocity model) (c) Comparison of the vertical stress component and S-wave velocity (black isolines) (d) Comparison of the horizontal stress component and the S-wave velocity (black isolines).

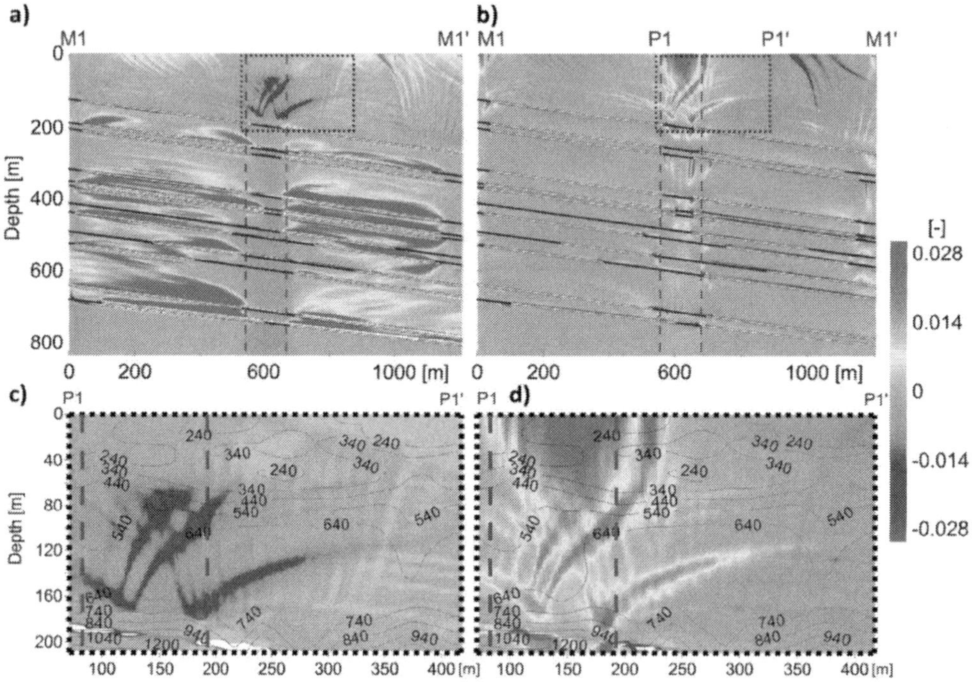

Fig. 6. (a) Vertical strain field, (b) Horizontal strain field (black dashed line indicates the range of S-wave velocity model) (c) Comparison of the vertical strain component and S-wave velocity (black isolines) (d) Comparison of the horizontal strain component and S-wave velocity (black isolines).

5 Conclusions

This article presents the results of a comparison between the numerical calculations of the stress and deformation state in a rock mass disturbed by multi-seam coal exploitation and an empirical model of the S-wave velocity obtained on the basis of the seismic interferometry method. Based on the outcomes of the research conducted, the following conclusions were drawn:

1. In the areas of increase of vertical and horizontal stress within the pillar, which consisted of coal seam remnants, the velocity of the S-wave is greater in comparison with the destressed zones of the rock mass.
2. In the areas of high values of vertical and horizontal strain within the top part of the pillar, the S-wave velocity decreases.
3. The limited range of the S-wave velocity model does not allow comprehensive imaging of the disturbed rock mass to be carried out.
4. The preparation of 3D models: numerical and empirical S-wave velocity would allow one to obtain more detailed results.
5. Presented methodology can be useful to study subsidence process and monitoring of bent layers.

This article was prepared as a result of the LOFRES Project No PBS1/A2/13/2013 performed within the 1st call of the Applied Research Programme co-financed by the National Centre for Research and Development in Poland. We thank the editor and two anonymous reviewers for valuable comments and discussions.

References

1. E. Popiołek, Z. Pilecki, *Evaluation of development possibilities of mining areas threatened with deformations using geophysical methods* (MEERI PAS, Kraków, 2005)
2. H. Marcak, Z. Pilecki, *Some geophysical and geomechanical remarks on recognition of sinkhole processes in post-mining areas*, Proc. of the 12th European Meeting of Environmental and Engineering Geophysics, Helsinki, Finland, 1–5 (2006)
3. Z. Pilecki, *The role of geophysical methods in the estimation of sinkhole threat in the post-mining areas of shallow exploitation in the Upper Silesian Coal Basin, Poland* Gospad. Surowcami. Min. **24**(3/1), 27–40 (2008)
4. S. Gibowicz, A. Kijko, *An Introduction to Mining Seismology* (Academic Press, New York, 1994)
5. A. J. Mendecki, *Seismic monitoring in mines* (Chapman and Hal, London, 1997)
6. A. Nur, J. Geophys. Res. **76**(8), 2022–2034 (1971) doi:10.1029/JB076i008p02022
7. Z. Szreder, Z. Pilecki, J. Kłosiński, *Comparison of Profiling Results of Attenuation and Velocity of Refracted P-wave in Coal-seam*, Proc. of the 14th European Meeting of Environmental and Engineering Geophysics, Kraków, Poland (2008)
8. J. Ślizowski, Z. Pilecki, K. Urbańczyk, E. Pilecka, L. Lankof, R. Czarny, Adv. High Energy Phys., 1–12 (2013) doi :10.1155/2013/461764
9. N. Hosseini, K. Oraee, K. Shahriar, K. Goshtasbi, J. Cent. South Univ. 19(8), 2297–2306 (2012) doi :10.1007/s11771-012-1275-z
10. Z. Pilecki, M. Laskowski, A. Hryciuk, J. Wróbel, E. Koziarz, E. Pilecka, R. Czarny, K. Krawiec, *Identification of gaso-geodynamic zones in the structure of copper ore deposits using geophysical methods*, CIM Journal 5(3), 194–202 (2014)

11. A. Cao, L. Dou, W. Cai, S. Gong, S. Liu, G. Jing, Int. J. Rock Mech. Min. Sci. 78, 1–9 (2015) doi: 10.1016/j.ijrmms.2015.05.001
12. G. Olivier, F. Brenguier, M. Campillo, R. Lynch, P. Roux, J. Geophys. Res. 80(3), KS11–KS25 (2015a) doi : 10.1190/geo2014-0299.1
13. G. Olivier, D. Hollis, F. Brenguier, A. Mordret, T. Lecocq, Proc. of the 79th EAGE Conference and Exhibition 2017, Paris, France (2017) doi: 10.3997/2214-4609.201700961
14. G. Olivier, F. Brenguier, M. Campillo, P. Roux, N. M. Shapiro, R. Lynch, Geophys. Res. Lett. 42, 1–9 (2015b) doi: 10.1002/2015GL065975
15. R. Czarny, H. Marcak, N. Nakata, Z. Pilecki, Z. Isakow, Pure Appl. Geophys. 173(6), 1907–1916, (2016) doi :10.1007/s00024-015-1234-3
16. A. Curtis, P. Gerstoft, H. Sato, R. Snieder, K. Wapenaar, The Leading Edge 25(9), 1082–1092 (2006) doi : 10.1190/1.2349814
17. B.H.G. Brady, E.T. Brown, *Rock mechanics: for underground mining* (Chapman & Hall, London, 1993)
18. M. A. Coultard, Geotechnical and Geological Engineering 17(3-4), 373–385 (1999) doi : 10.1023/A:1008951216602
19. L. Jing, J. A. Hudson, Int. J. Rock Mech. Min. 39(4), 409–427 (2002) doi : 10.1016/S1365-1609(02)00065-5
20. M. Kwaśniewski, J. Wang, *Numerical analysis of mining-induced deformations of the strata*, Proc. of The School of Underground Mining '96, Kraków, Poland, 153–187 (in Polish) (1996)
21. J. Kook, C. Scior, P. Fischer, M. Hegemann, *Subsidence predictionfor multiple seam extraction under consideration of time effects by the use of geomechanical numerical models,* Proc. of the 27th International Conference on Ground Control in Mining, WV, USA, 143–147 (2008)
22. J. Han, H. W. Zhang, P. Zhang, J. Zhang, M. Li, *Overmining in Kailuan, China mining area*, Proc. of the 28th International Conference on Ground Control in Mining, WV, USA, 305–311 (2009)
23. Z. Pilecki, *Modelowanie numeryczne pola naprężenia w górotworze naruszonym wielopokładową eksploatacją węgla kamiennego w warunkach silnego zagrożenia sejsmicznego*, Zeszyty Naukowe - Bulletin of the Mineral and Energy Economy Research Institute of the Polish Academy of Sciences 80, 93–102 (in Polish) (2011)
24. M. Wesołowski, *Modelowanie numeryczne wpływu zaszłości eksploatacyjnych na deformacje terenu górniczego na przykładzie kopalni „Chwałowice"*, Przegląd Górniczy 69(6), 27–35 (in Polish) (2013)
25. M. Zhang, H. Shimada, T. Sasaoka, K. Matsui, L. Dou, Environ. Earth Sci. 72(3), 629–643 (2014) doi :10.1007/s12665-013-2985-8
26. W. Sui, Y. Hang, L. Ma, Z. Wu, Y. Zhou, G. Long, L. Wei, B. Eng. Geol. Environ. 74(3), 1019-1035 (2015) doi :10.1007/s10064-014-0674-9
27. G.A. Zhu, L.M. Dou, Z.L. Li, W. Cai, Y. Kong; J. Li, Arab. J. Geosci. 9, 365 (2016) doi :10.1007/s12517-016-2356-3
28. G.D. Bensen, M.H. Ritzwoller, M.P. Barmin, A.L. Levshin, F. Lin, M.P. Moschetti, N.M. Shapiro, Y. Yang, Geophys. J. Int. 169(3), 1239–1260 (2007) doi: 10.1111/j.1365-246X.2007.03374.x
29. J. Xia, R.D. Miller, C.B. Park, Geophysics 64(3), 691–700 (1999) doi: 10.1190/1.1444578

30. Z. Isakow, Z. Pilecki, A. Kubańska, *System LOFRES sejsmiki pasywnej z wykorzystaniem szumu sejsmicznego* (Instytut Technik Innowacyjnych EMAG, Katowice, in Polish, 2015)

Numerical simulation of pore pressure changes in levee under flood conditions

Jacek Stanisz[1*], *Aleksandra* Borecka[1], *Zenon* Pilecki[2] and *Robert* Kaczmarczyk[1]

[1]AGH University of Science and Technology, 30-059 Krakow, Poland
[2]Mineral and Energy Economy Research Institute, Polish Academy of Sciences, 31-261 Krakow, Poland

Abstract. The article discusses the potential for using numerical simulation to assess the development of deformation and pore pressure changes in a levee as a result of the increase and decrease of the flood wave. The simulation made in FLAC 2D did not take into account the filter-erosion deformation associated with seepage in the levee. The simulations were carried out for a field experimental storage consisting of two combined levees, which was constructed with the help of homogeneous cohesive materials with different filtration coefficients. Calculated and measured pore pressure changes were analysed at 4 monitoring points. The water level was increased to 4 m in 96 hours and decreased in 120 hours. The characteristics of the calculated and measured pore pressure changes over time were similar. The maximum values of the calculated and measured pore pressure were almost identical. The only differences were the greater delay of the experimental levee response to changes in water level increase compared to the response of the numerical model. These differences were probably related to filtering-erosion effects during seepage in the levee.

1 Introduction

Levees are a basic element of the flood protection system [1]. In Poland their total length is about 8 500.0 km. Nearly 60% of them are over 40 years old, only 20% have been in use for 20 years [2].

Deformation of the levee results from the hydraulic pressure of water, water filtering through the levee and its subsoil, and the hydrating and drying of the levee material [3]. This process is intensified under the influence of an increased level of water during a flood. Under conditions of change in water pressure, the development of the deformation process often leads to the failure of the levee [4-6].

The article presents the results of numerical simulation of the behaviour of the levee under an increase and decrease of the hydraulic pressure for water levels up to 4 m in height. We analysed the pore pressure changes in the levee. The results of numerical calculations were compared with the results of field measurements inside of an experimental water storage facility, which was built under the ISMOP Project (IT System

* Corresponding author: jstanisz@agh.edu.pl

DOI: 10.1051/e3sconf/20172403002

of Levee Monitoring) [7-8]. The study is based also on results a few experiments conducted under the project [9-12].

2 Characteristics of the research area

The research area is located in Czernichów village near the Vistula river in southern Poland, about 20 km in west of Cracow. The water storage of 208 x 58 x 4.5 m (L x W x H) forms two sections of levees joined to each other (Fig. 1). The volume of the structure is about 28.500 m^3. The levee slopes have variable inclination - the waterside is 1:2 and 1:2.5, and the landside is 1:2.

The levee on the eastern side was built of silty clay, and on the west side of silty sand with filtration coefficients of $1.83 \cdot 10^{-5}$ and $5.24 \cdot 10^{-5}$ m/s respectively. The subsoil is formed of silty clay and clayey sand, sands and gravels. Ground-water conditions were identified by boreholes and dynamic and static sounding [13]. The levee in contact with subsoil was isolated by impermeable bentomate.

In the field measurements, the influence of the increase and decrease of the water level on the levee's behaviour was simulated [14-16]. The changes of water pressure were measured by pore pressure sensors.

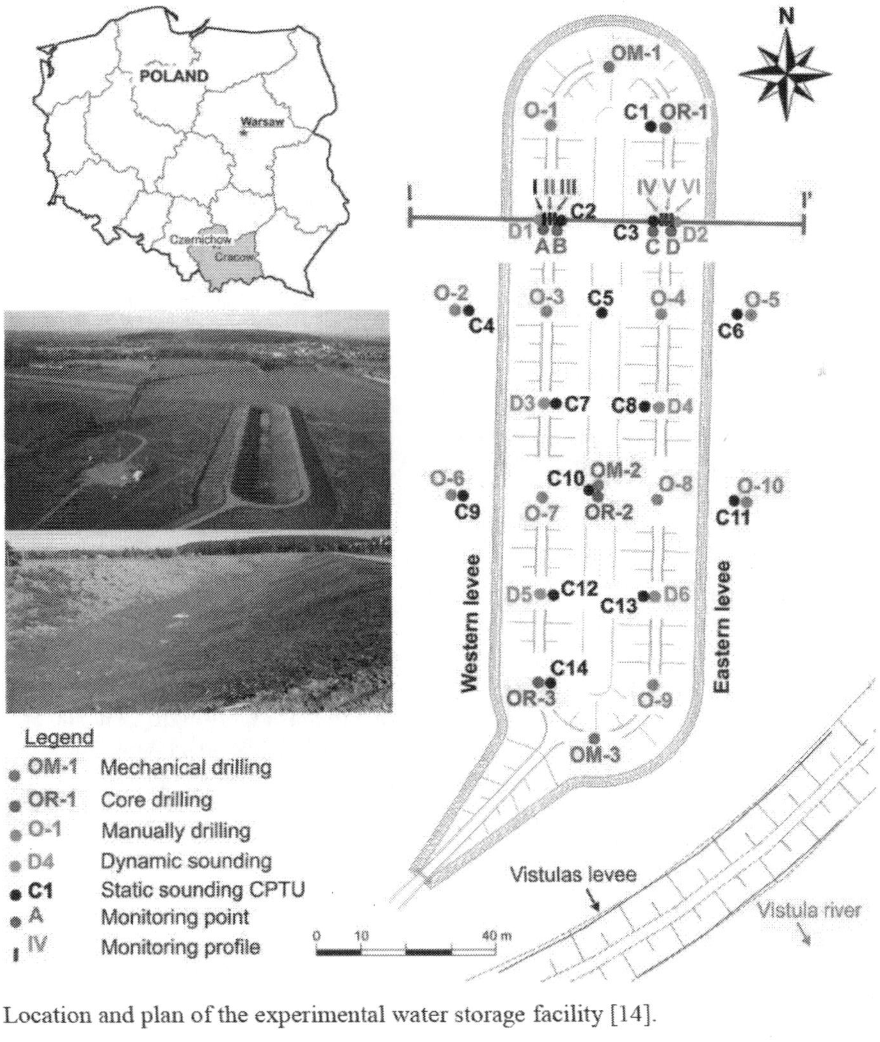

Fig. 1. Location and plan of the experimental water storage facility [14].

DOI: 10.1051/e3sconf/20172403002

3 Methodology

3.1 Calculation scheme

A schematic diagram of the calculation procedure is shown in Figure 2. In calculations we assumed a continuous increase of water level up to 4 m during 96 hours and decrease within 120 hours. The numerical simulations were made in plane state of strain with the help of FLAC 2D v. 7.0 code. The pore pressure changes were analysed in the selected cross section I – I' and in 4 monitoring points located in the lower part of the levees, shown in Figure 1. Points A and B were placed in an levee on the eastern side of the water storage facility and points C and D in an levee on the western side.

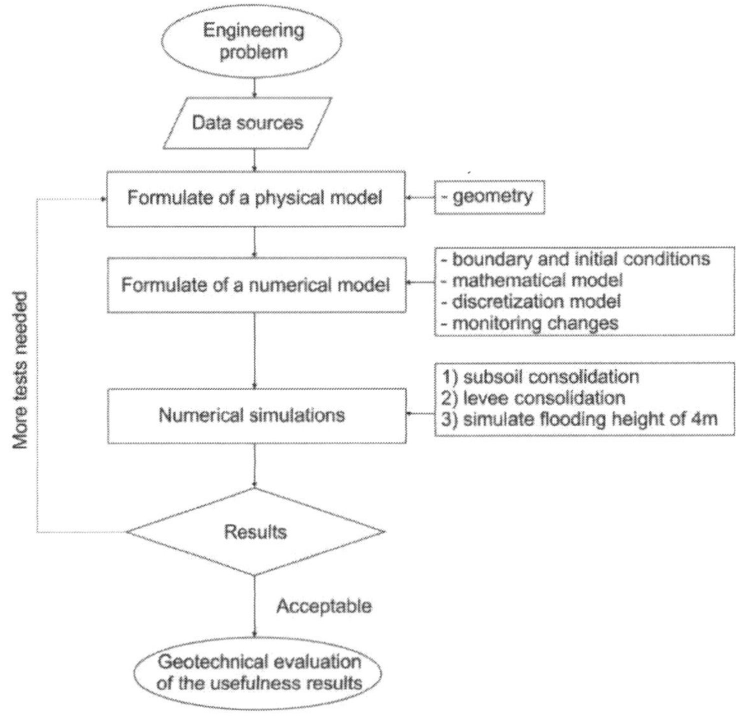

Fig. 2. A schematic diagram of a numerical simulation [17-18].

3.2 Physical model

The physical model was prepared along cross-section I-I' running through the central part of the experimental storage (Fig. 3). It is assumed that the vertical frame is 11 m high and the horizontal frame is 90 m long. There are 6 geotechnical layers with material properties listed in Table 1. The first layer corresponds to the levee material from the eastern or western side of the experimental storage. The remaining 5 layers are located in the subsoil of the storage.

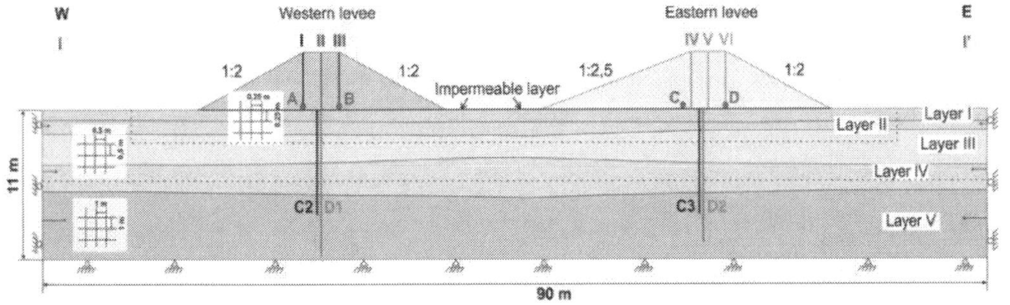

Fig. 3. Physical model of the water storage with subsoil along cross-section I-I'.

Table 1. Material constants of levees and subsoil layers.

Number of layer	Bulk density [Mg/m³]	Cohesion [kPa]	Angle of internal friction [deg]	Bulk modulus [kPa]	Shear modulus [kPa]	Filtration coefficient [m/s]
Eastern levee	1.9	12.5	30.04	$8.53e^{3}$	$3.27e^{3}$	$1.83e^{-5}$
Western levee	1.96	15.43	35.2	$7.25e^{3}$	$3.35e^{3}$	$5.24e^{-5}$
I	2.1	10.3	32.9	$7.25e^{3}$	$3.43e^{3}$	$1.52e^{-5}$
II	1.89	13.7	22.5	$16.2e^{3}$	$6.63e^{3}$	$1.35e^{-5}$
III	1.85	9.5	35.4	$16.2e^{5}$	$6.63e^{4}$	$1.32e^{-6}$
IV	1.9	9.3	39.4	$16.2e^{5}$	$6.63e^{4}$	$1.32e^{-5}$
V	2.05	7.2	40.1	$36.8e^{5}$	$21.8e^{4}$	$5.5e^{-5}$

3.3 Calculation model

In calculations, we assumed a elastic-plastic mechanical model with use of Mohr-Coulomb strength criterion. Hydrostatic pressure was applied on the waterside of the levees. Water flow was allowed on the landside of the levees. The model was discretised with a regular grid of 0.25 m for the levee, 0.5 m for layer I-III and 1 m for layers IV and V. The effect of the increase in water level was analysed: (1) up to 2 m in 48 hours, (2) up to 4 m in 96 hours and (3) with return to the initial situation in 120 hours.

4 Results and analysis

The results of numerical simulations in the form of isoline changes of pore pressure, shear stress and horizontal displacement are shown in Figures 4-6.

The increase of the water level up to 2 m resulted in an increase of pore pressure up to 20 kPa at the bottom part of the levee (Fig. 4a). For the assumed duration of waterlogging, the shear stresses reached values of -15 to 20 kPa on the landside of the experimental storage (Fig. 4b). The levee began to deform and the horizontal displacement reached 7 mm (Fig. 4c). In this case, the maximum water flow velocity in the levees was 0.33 m/d.

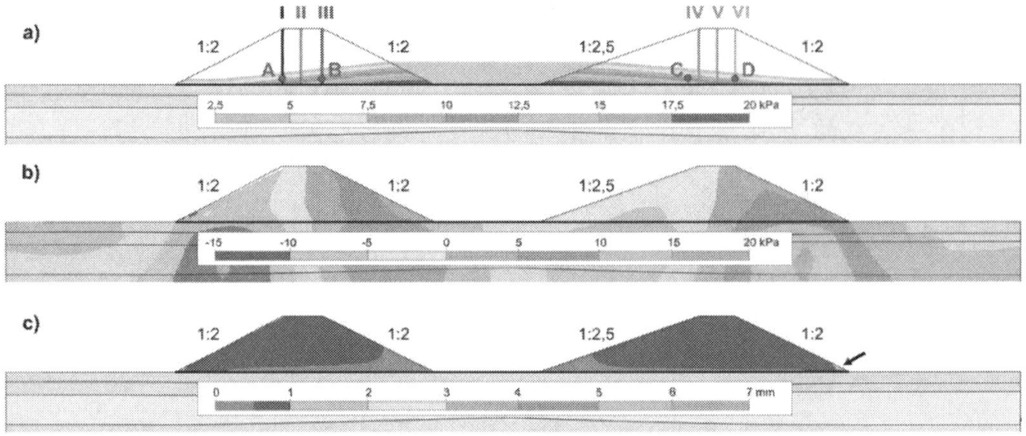

Fig. 4. Pore pressure field (a), shear stress (b) and horizontal displacement (c) inside the levee for an increase in water level to a height of 2 m. The black arrow indicates the zone of maximum displacement (7 mm).

In these conditions, the water level increased to 4 m, the pore pressure increased to 40 kPa and the deformation process increased in the bottom part of the levees (Fig. 5a). In the levees a considerable decrease in pore pressure value to several kPa was observed, presumably due to an increase in the velocity of water flow to 3.5 m/d. On the landside of the levees, horizontal displacement reached a value of 14 mm (Fig. 5c). In the surface part of the waterside of the levees, an increase in shear stress of up to 40 kPa was observed (Fig. 5b).

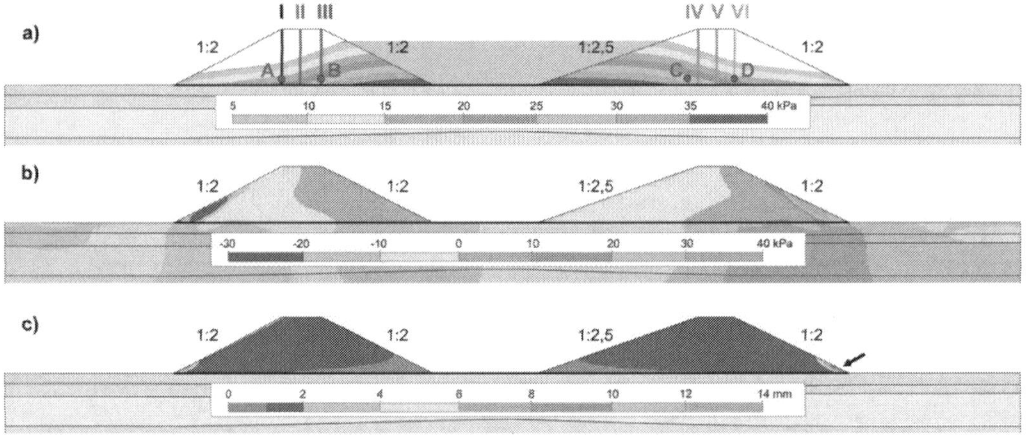

Fig. 5. Pore pressure field (a), shear stress (b) and horizontal displacement (c) inside the levee for an increase in water level to a height of 4 m. The black arrow indicates the zone of maximum displacement (14 mm).

As a result of a water level decrease to 1 m, the pore pressure reached a value of 18 kPa (Fig. 6a). Under these conditions, the maximum displacement achieved was 3 mm (Fig. 6c), and the water flow velocity inside the levees decreased to about 0.13 m/d (Fig. 6b).

 DOI: 10.1051/e3sconf/20172403002

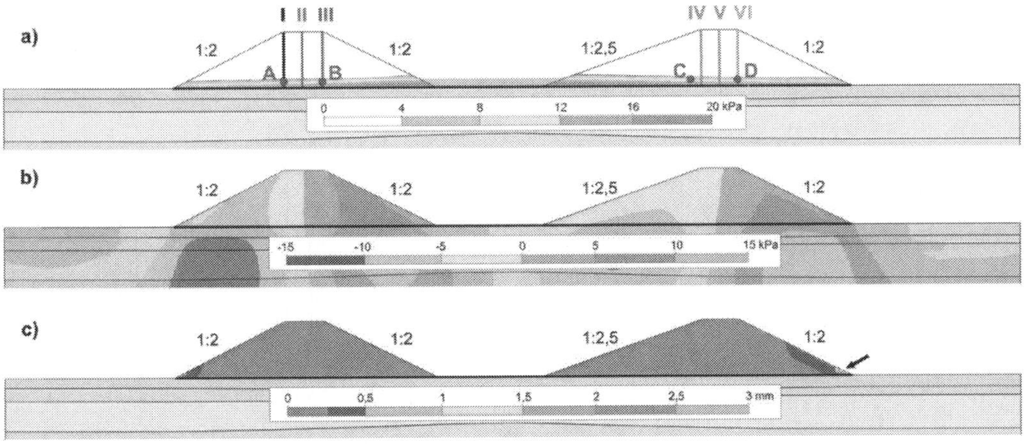

Fig. 6. Pore pressure field (a), shear stress (b) and horizontal displacement (c) inside the levee for a decrease in water level to a height of 1 m. The black arrow indicates the zone of maximum displacement (3 mm).

In general, the pore pressure in the numerical model increases and decreases approximately in accordance with changes in water level (Fig. 7). By comparing the pore pressure changes at the four monitoring points A, B, C and D in the numerical model and in the field experimental storage, we observed that the characteristics of these changes is similar. The maximum values of pore pressure calculated and measured are almost identical. The differences only occurred with the late response of the experimental levee compared to the response of the levee in the numerical model (Fig. 7). Presumably, these differences are related to filter-erosion effects during water flow through the experimental levee. Under conditions of water level decrease, the calculated and measured pore pressure values were similar at all monitoring points.

It should be emphasised that the short time of increase and decrease of the water level limited the development of deformation in the levee. Under the conditions of the field experiment, the asymmetric levee built of silty sand was more susceptible to deformation, with higher pore pressure values of approximately 5-7 kPa (Fig. 8).

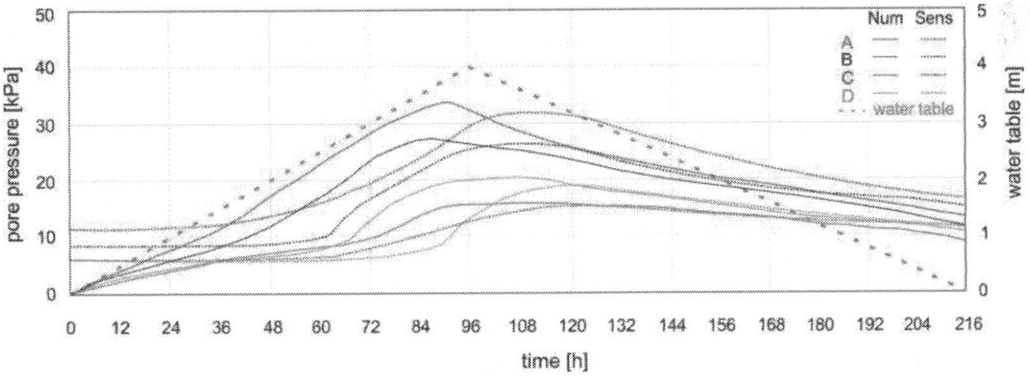

Fig. 7. The changes of pore pressure at four monitoring points (A-D) during a 4 m increase in the height of the water level in the numerical model (Num) and field experiment (Sens).

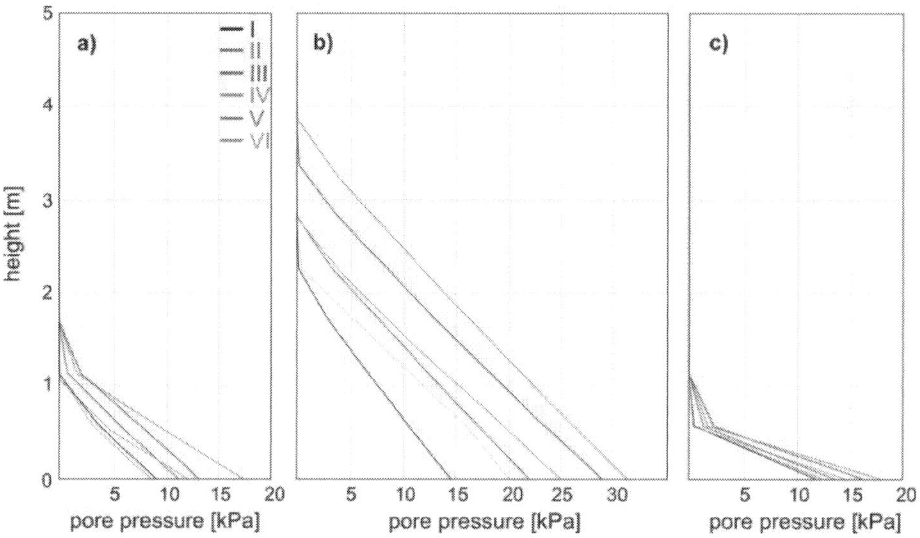

Fig. 8. Changes of pore pressure in sections I-VI for an increase in water level up to 2 m (a) 4 m (b) and its decrease to 1 m (c) in field measurements.

5 Conclusions

The results of a numerical simulation of the influence of hydraulic pressure on the levee allow us to draw the following conclusions:

1. Under conditions of water level increase, the changes of pore pressure in time and its maximum values in the numerical model of the levee and in the field experimental levee are similar. The differences in behaviour of both levees only occurred with the late response of the experimental levee comparing to the response of the levee in the numerical model. Presumably, these differences are related to filter-erosion effects during water flow through the experimental levee.
2. Under conditions of water level decrease, calculated and measured pore pressure changes in time were similar in all monitoring points.
3. For the increase in water level up to 4 m height in the numerical model, the greatest displacements occurred in the bottom part of the levee on its landside. Water outflow was observed on the surface of this slope in the field experimental levee. The calculated shear stress was 30-40 kPa. Horizontal displacement was increased to 14 mm.
4. The results obtained show the significant influence of hydraulic pressure caused by the increase in water level on pore pressure changes and possible development of the levee damage process [19].

The research presented in this paper was supported by the National Centre for Research and Development (NCBiR) under Grant No. PBS1/B9/18/2013

References

1. J. Stanisz, A. Borecka, A. Leśniak, K. Zieliński, *Wybrane systemy monitorujące obwałowania przeciwpowodziowe*, Przegląd Geologiczny **62,** 699-703 (2014)
2. J. Filaber, B. Kosowski, A. Borecka, *Flood protection in crisis management system* (Texter Warsaw, 2016)

3. A. J. Lobbestael, A. Athanasopoulos-Zekkos, J. Colley, Math. Probl. Eng., **2013**, 1-13 (2013) doi 10.1155/2013/893602
4. Z. Pilecki, *Rozpoznanie metodą sejsmiczną stanu podłoża obwałowań przeciwpowodziowych*, Bezpieczeństwo Pracy i Ochrona Środowiska w Górnictwie Miesięcznik Wyższego Urzędu Górniczego **5**, 37-39 (2004)
5. U. el Shamy, F. Aydin, J. Geotech. Geoenviron. **134**, 9, 1385-1398 (2008) doi 10.1061/(ASCE)1090-0241(2008)134:9(1385)
6. T. Zydroń, A. Gruchot, *Wpływ opadu deszczu i erozji bocznej na stateczność na przykładzie osuwiska w Bieśniku (Beskid Niski).* Acta Scientiarum Polonorum, Seria Architectura **15**, 3, 149-161 (2016)
7. J. Stanisz, K. Korzec, A. Borecka, *ISMOP Project (IT System of Levee Monitoring) as an example of integrated monitoring of levee*, Geology, Geophysics & Environmental **41**, 137-139 (2015)
8. A. Borecka, J. Stanisz, K. Sekuła, A. Leśniak, *Smart levee in Poland*, Abstract Book 3th FLOODRisk, **3**, 258 (2016)
9. B. Bukowska-Belniak, M. Dwornik, A. Pięta, A. Leśniak, *A 2D model of temperature changes in experimental embankment*, Measurement, Automation, Monitoring **61**, 6, 233-236 (2015)
10. K. Krawiec, B. Bukowska-Belniak, A. Leśniak, D. Kessler, E3S Web Conf. **7** (2016) doi: 10.1051/e3sconf/20160703018
11. M. Dwornik, K. Krawiec, A. Franczyk, A. Leśniak, E3S Web Conf. **7** (2016) doi: 10.1051/e3sconf/20160703018
12. A. Pięta, K. Krawiec, Procedia Comput. Sci. **51**, 2668-2677 (2015) doi: 10.1016/j.procs.2015.05.377
13. J. Mościcki, G. Bania, M. Ćwiklik, A. Borecka, *DC resisitivity studies of shallow geology in the vicinity of Vistula river flood bank in Czernichów village (near Kraków in Poland)*, Studia Geotechnica et Mechanica **36**, 1, 63-70 (2014)
14. M. Chuchro, A. Franczyk, B. Bukowska-Belniak, A. Leśniak, *Reducing flood risk using computer system for monitoring river embankments*, Environmental Protection and Natural Resources **28**, 3, 73, 11-16 (2017)
15. A. Pięta, M. Chuchro, M. Lupa, A. Piórkowski, A. Leśniak, Lect. Notes. Comput. SC. **8838**, 514-525 (2014) doi: 10.1007/978-3-662-45237-0_47
16. A. Franczyk, M. Dwornik, A. Leśniak, *Numerical modelling of the impact of flood wave cyclicality on the stability of levees*, Abstract Book 3th FLOODRisk, **3**, 1-6 (2016)
17. J. Stanisz, Z. Pilecki, H. Woźniak, *Wybrane zagadnienia numerycznej analizy stateczności osuwiska w Swoszowicach,* Technika Poszukiwań Geologicznych. Geotermia, Zrównoważony Rozwój **2**, 77-89 (2012)
18. M. Dyba, Z. Pilecki, *Wpływ sposoby zawodnienia na ciśnienie porowe i naprężenie efektywne w obliczeniach numerycznych programem FLAC 2D*, Przegląd Górniczy **7**, 25-31 (2012)
19. P. Popielski, B. Smoliński, A. Kasprzak, *Ocena stanu obiektów hydrotechnicznych na podstawie wyników interdyscyplinarnych programów badań i symulacji numerycznych*, Czasopismo Inżynierii Lądowej, Środowiska i Architektury **63**, 25-32 (2016)

AUTHOR INDEX